THE BATTLEGROUNDS OF BIO-SCIENCE

Cross-Examining the Experts on:
Evolutionary Psychology
Race, Intelligence, & Genetics
Population, Environment, & Cloning

Interviews of:
Richard Dawkins,
Lionel Tiger & Robin Fox,
E. O. Wilson,
Donald Johanson,
Charles Murray, Robert Sternberg,
Julian Simon, Garrett Hardin,
Richard Seed

Plus the Author's
"Quick & Dirty Guides" & Reviews

Frank Miele
Senior Editor,
SKEPTIC Magazine

ISBN: 1-4033-6029-4 (e-book)
ISBN: 1-4033-6030-8 (Paperback)

This book is printed on acid free paper.

1stBooks – rev. 11/15/02

"Charles Darwin will eventually be seen as a far more influential figure in the history of human thought than either Jesus Christ or Mohammed."

—James D. Watson,
Co-winner of the 1962 Nobel Prize
in Physiology or Medicine
for the discovery of the DNA Double Helix

To the Memory of my Bull Terriers,
Patrick, then *Jethro*;
And now to my Great Dane, *Payce*—
True Scions of the House of *Argus*,
And, Successively, This Man's Best Friend

ACKNOWLEDGMENTS

I would like thank Michael Shermer, publisher and Editor-in-Chief of *Skeptic* Magazine for first inviting me to write for *Skeptic,* and then for granting permission to reprint these interviews and articles, and to the entire *Skeptic* staff, especially Art Director Pat Linse and Tanja Sterrmann for their work on the graphic and text files. I would also like to thank Profs. Marina Butovskaya and Valery Tishkov of Russian Academy of Sciences and Dr. Frank Salter of the Max Planck Institute for inviting me to present a summary of my work at the Evolution, Behavior, and Society Conference on Human Ethology held at Zvenigorod, Russia, 19-26 June 2001 and all their students for the interest and appreciation they showed in my work. There are many others who have assisted and encouraged me on this project, without whom it would never have been accomplished. Finally, I must thank all those who so graciously gave of their valuable time to be interviewed.

TABLE OF CONTENTS

PREFACE

The start of the new millennium has seen the dawn of Charles Darwin's dream — the synthesis of the physical, biological, and social sciences, what E. O. Wilson (who author Tom Wolfe dubbed "Darwin II") has called "consilience." But to a vast portion of the American population, especially the Religious Right, human evolution itself is a dangerous heresy. They fear that emerging neo-Darwinian disciplines such as Sociobiology and Evolutionary Psychology are heading us to a world where the "there are no absolutes and anything goes." Many on the Radical Left, including postmodernists, deconstructionists, militant feminists, and unrepentant Marxists fear the explosion of knowledge on the genetic basis of disease and behavior is leading to a newer, seemingly more acceptable Social Darwinism. They fear the rise of a dangerous form of corporate-controlled biotechnology and eugenics where "the strong do as they will and the weak suffer what they must." What all these very strange bedfellows share is that they all subscribe to the dogma that a certain collection of written words (whether attributed to God, Allah, Karl Marx, or "The Womyn's Collective,") must take priority over observations of the real world.

The dawn of the Darwinian dream also marks the dawn of the Darwinian dilemma — has today's knowledge of genetic basis of life, disease, and intelligence, and the evolutionary origins of parental care, violence, and race hatred outstripped the ability of our evolved brains to process that information wisely. Those are among the questions I explored in my Skeptic Magazine interviews with some of today's most controversial and far thinking scholars. Under skeptical cross examination, their opposing points of view map the dimly-seen borderlands where science, ethics, and politics will meet in the next century. The result will be a new definition of life, of intelligence, and of no less than what it is means to be human.

Also included are my *Skeptic* Magazine "Quick and Dirty" Guides to emerging disciplines such as Evolutionary Psychology, Darwinian Anthropology, Human Behavioral Genetics, and Chaos and Complexity Theory and reviews of some of the most important, if not necessarily best known or best selling books about them.

It is my hope that the interviews, essays, and reviews collected here in *The Battlegrounds of Bio-Science* will provide general readers, college and university students, and media figures with in-depth analyses of the conflicting visions of some of today's most thought provoking scholars and best-selling science writers, their personalities, and their replies to critics (including other interviewees). College classes, graduate seminars, and reading discussion groups should find many topics for debate, term papers, or their own further research. Journalists and media figures may also find them useful as 'quick start' guides in researching articles and preparing for their own interviews.

The Battlegrounds of Bio-Science is divided into four parts:

Part I: The Evolutionary Psychology Debates asks whether there is an underlying human nature, produced by Darwinian evolution and encapsulated in modules in the human brain, or whether evolutionary explanations are at best naive and at worst a smoke screen for political oppression.

Part I begins with "The Quick and Dirty Guide to Evolutionary Psychology and the Nature of Human Nature." It provides a brief, and at times humorous introduction to the emerging discipline of evolutionary psychology, which sets the stage for the interviews that follow. Chapter 2 is an interview of Darwin's Dangerous Disciple — Richard Dawkins, author of *The Selfish Gene*. He describes a mechanistic universe of selfish genes driven by natural selection, devoid of design, purpose, good or evil. Dawkins cautions against the Naturalistic Fallacy of mistaking "Is" for "Ought," and argues that religion is a mind virus. Chapter 3 is an interview of anthropologists Lionel Tiger and Robin Fox, who look back on how the social sciences have changed since 1971 when they wrote their ground-breaking book, *The Imperial Animal*. Tiger then criticizes the real danger of evolutionists, such as Richard Dawkins in the preceding interview, "looking at religion in a mocking sense and concentrating on its negative features." Fox also argues, contra Dawkins, that philosopher David Hume did not say that one couldn't logically go from Is to Ought, but rather made such jumps all the time, and so did Darwin.

The fourth chapter in Part I is an interview of Edward O. Wilson, the father of sociobiology. He discusses his book, *Consilience*, where he argues that because of the recent explosion of knowledge from the Human Genome Project and the Decade of the Brain we stand on the verge of reaching consilience — the age-old goal of unifying knowledge from the sciences and the humanities into one grand philosophy. I ask Wilson to respond to the criticisms of philosopher Richard Rorty and of the late biologist Stephen Jay Gould that science, art, and religion are (to use Gould's Neologism) Non-Overlapping Magisteria (NOMA) that can never be united. "Can we unify knowledge through the consilient approach or should we compartmentalize it and lead our lives as if we were split-brained?" Wilson's retorts that "These strong, even emotional objections ... are what I call the secular intellectuals' white flag,... politically correct in an almost all-encompassing sense."

Part I concludes with review essay of four important books on the evolution of human nature, cooperation, and violence, which adopts a somewhat jocular tone in citing rock 'n' roll titles and lyrics to illustrate its points. It also takes a skeptical look at the skeptical movement's treatment of the Darwinian revival.

Part II: The Origins and Diversity Debates explores human evolution and human differences. The first chapter, Chapter 6, sets the stage by using the figure of Caliban in Shakespeare's play, *The Tempest* ("A devil, a born devil, on whose nature Nurture can never stick;"), to trace the tempestuous history of anthropology — especially the contentious issues of race and cultural and evolutionary progress. Chapter 7 then asks whether evolution is just another origin myth, as religious fundamentalist and literary deconstructionists (a rather strange set of bedfellows) argue. It reviews the origin and function of origin myths from those of hunter gatherers to those contained in literary epics from *The Iliad*, *The Odyssey*, and *Exodus* to novels/movies like *Exodus*, *The Godfather*, and *The Winning of the West*.

Chapter 7 is an interview of paleoanthropologist Donald Johanson, discoverer of the famous Lucy fossil. He tells how a front page news photo of a biology teacher in a fundamentalist school holding up a bumper sticker reading, "Darwin is dead and he ain't coming back!" got his juices flowing. So much so that he wrote a letter to the editor declaring, "Isn't it a shame that ... we're still teaching ignorance to our children. Sir Isaac Newton is dead, but gravity ain't going away! Maybe we should have that as a bumper sticker!"

Turning to the ever controversial topic of race and brain size, I ask Johanson, "do the anatomical differences in living humans ... mean something behaviorally, or [if not] how do you know those differences meant something ... in the evolutionary record? If someone suddenly cloned a Neanderthal, or a *Homo erectus*, or an *Australopithecus afarensis*, how do we know that they would be behaviorally different from people alive today?" "You've really sent up a balloon here," Johanson replies. ... "There are clearly populational differences... [B]ut ... whether you are a native Australian or a native Laplander, what's going on inside the braincase is essentially the same. They're capable of all the same sorts of behaviors as any living member of *Homo sapiens*."

Part II concludes with "The Quick and Dirty Guide to Chaos and Complexity Theory." It explores the questions of progress and contingency in evolution (as indicated by increasing brain size in different phylogenetic lines) and the use of counterfactuals in history.

Part III: The Race-IQ-Genetics Debates explores whether IQ tests truly measure "intelligence;" If so, does it have a genetic component; and the most controversial issue of all – what causes the 15-point difference in average IQ between Blacks and Whites in the U.S.?

Part III begins with an interview of Charles Murray, co-author of the most controversial best-selling book, *The Bell Curve*. Murray and his co-author, (Harvard psychologist Richard Herrnstein, who died shortly before the book's publication) argued that IQ is real; it is a good predictor not only of academic grades but of job performance, income, crime, and illegitimacy; it is between 40 and 80% the result of genetic, not socioeconomic factors; it will become increasingly important as society becomes more technological; and that at least some part of the Black-White average IQ difference is genetic.

Asked to defend the thesis of his book against the harsh criticism in received in the media as being a work of "pseudoscience," Murray cites the results of a 1991 survey of expert opinion among behavior geneticists and psychometricians (i.e., IQ experts) who agreed that IQ was real, highly predictive, had a significant genetic basis, and at least some of the black-white IQ difference probably had a genetic cause. Murray declares "The science in *The Bell Curve*, Murray declares, "is very much in the middle of the mainstream." and then adds, "[T]here has been a collective intellectual cowardice about understanding the role of intelligence in understanding social problems."

In the following chapter, Robert Sternberg, editor of *The Encyclopedia of Intelligence*, accuses Herrnstein and Murray of erroneously "inviting their readers to believe" that if intelligence is highly heritable it cannot be changed by environmental intervention. Sternberg states that there is always an interaction between heredity and environment, that we just don't know what causes the Black-White difference in average IQ, and concludes that the cognitive stratification "revealed" by *The Bell Curve* is actually "an invention of our society."

The next chapter is a review essay of five important books on human behavior genetics that illustrate the mismatch between what popular media accounts on the one hand, and the professional literature on the other, have to say on the subjects of IQ, race, and genetics. What is presented in the popular media as "debunking" of the genetic viewpoint turns out in the technical literature to be "qualifications, caveats, and concessions." In the words of cognitive psychologist Earl Hunt, since the Nature versus Nurture debate has now become a "stomping match between Godzilla [Nature] and Bambi [Nurture]," it is time to move on to discover "the causal pathway… between genes and behavior." Part III concludes with a review of *Entwined Lives: Twins and What They Tell Us about Human Behavior* by Nancy Segal, an expert on the subject and a twin herself. Segal book entwines the borderland emerging disciplines of evolutionary psychology, behavior genetics, and developmental genetics. The real questions in behavioral research today are how nature works through nurture and how nurture selects nature, and how all these factors interact during development.

Part IV: The Population, Environment, and Cloning Debates explores takes a planetary focus. The "Quick and Dirty Guide to Population and Environment" sets the stage for the debate between ecologists, who claim human overpopulation is destroying the Earth's environment and we are quickly using up the supply of raw materials, and economists, who say, "Show me the money!" — commodities are getting cheaper, and life is getting better for more and more people. Text and diagrams trace the growth of population over the course of human evolution, the effect of industrialization on population growth, varying estimates of how many people the earth can support (and how wrong some of these estimates have been), and the changes in temperature and carbon dioxide levels.

The next chapter is an interview of Garrett Hardin, author of the famous essay, "The Tragedy of the Commons." According to Hardin, the earth already far more people that it can sustain and we are rapidly and perhaps irreparably destroying our environment. "Economic analysis," says Hardin, "is poorly fitted to deal with the future. ... [Economists] just deal with one human and then another and they constantly ... forget about the environment."

The Hardin interview is followed immediately by one of the late economist, Julian Simon. In 1980 ecologist and population control advocate Paul Ehrlich predicted that increases in population would produce an increased demand and commodity prices would sky rocket for the next 10 years. Simon, arguing that demand would spur exploration and provide an incentive to create substitutes, challenged Ehrlich to a bet. By the end of the decade, Simon had won hands down. Simon then explains why he later refused to bet Ehrlich and on an alternate list of 15 items of material human welfare such as life expectancy, leisure time, and purchasing power, which he characterizes as "switch and bait." Simon replies, "No matter how many 'statements' and 'warnings to the world' ecologists and biologists issue, they are not competent to answer questions about "human economic consequences." Only economists and statisticians can. Simon counters, "Pick any measure of material human welfare...pick any country... 10 years in the future... it will show improvement."

Part IV, and the book, conclude with two chapters on the controversial subject of cloning. First, "The Quick and Dirty Guide to Cloning" sets the stage by describing the scientific questions and ethical issues involved. It is followed by an interview of Richard Seed who explains how he is at work to clone the first human – himself.

The interviews, articles, and reviews collected here in *The Battlegrounds of Bio-Science* are reproduced exactly as they appeared in the various issues of Skeptic Magazine, other than for reformatting and correcting minor typographical errors in the originals.

Part I

The Evolutionary Psychology Debates

Frank Miele

CHAPTER 1

A QUICK & DIRTY GUIDE TO EVOLUTIONARY PSYCHOLOGY & THE NATURE OF HUMAN NATURE

THE (IM)MORAL ANIMAL

Is "the fault, dear Brutus, not in our stars but in ourselves?" In our genes? Or in our jeans? Why do some "bestride the narrow world like a Colossus" while other "petty men [most of us] peep about to find ourselves dishonorable graves"? Are not men, as Shakespeare suggested in *Julius Caesar*, at least sometimes "masters of their fates"? Or, as Jack Nicholson's "average horny little devil" asks about the differences between men and women, in the film version of Updike's *The Witches of Eastwick*:

> Do you think God knew what he was doing...or do you think it was just another of his minor mistakes—like tidal waves, earthquakes, floods....When we make mistakes, they call it evil; God makes mistakes, they call it nature.
>
> A mistake? Or did he do it on purpose? Because if it's a mistake, maybe we can do something about it—find a cure; invent a vaccine; build up our immune system.

3

Figure 1.0

Throughout most of human history, the answers to these questions have come from myth or literature. Starting with the Enlightenment, however, the answers have usually been couched in the allegedly "objective findings" of either history or science. Since the end of World War II, the "standard model of social science," as summarized by Robert Wright in his very readable introduction to evolutionary psychology, skeptically (if not cynically) titled *The Moral Animal*, has held that "the uniquely malleable human mind, together with the unique force of culture, has severed our behavior from its evolutionary roots;...[and] there is no inherent human nature driving events...our essential nature is to be driven" (1994, p. 5).

For example, Emile Durkheim, the patriarch of modern sociology, referred to human nature as "merely the indeterminate material that the social factor molds and transforms." He argued that even such deeply felt emotions as sexual jealousy, a father's love of his child, or the child's love of the father are "far from being inherent in human nature." Robert Lowie, a founding father of American cultural anthropology, argued that "the principles of psychology are as incapable of accounting for the phenomena of culture as is gravitation to account for architectural styles." Ruth Benedict, one of the founding mothers of American anthropology, and a crusader against the theory of racial differences (which was the norm in pre-World War II days), wrote that "we must accept all the implications of our human inheritance, one of the most important of which is the small scope of biologically transmitted behavior, and the enormous role of the cultural process of transmission of tradition." (All quotes from Wright, 1994.) B. F. Skinner's behaviorist psychology, dominant in American psychology in the 1950s and 1960s, was built on the bedrock assumption that all human and animal behavior could be accounted for in terms of rewards and punishments.

To all of this, evolutionary psychologists reply with all the gusto of *Saturday Night Live's* Wayne and Garth, *"Not!"* Human nature is real, it is important, and it isn't going to go away. Here is a sampling of the sorts of questions evolutionary psychologists ask and attempt to answer:

- Are we all naturally the same or naturally different?
- Is our mind all of one piece or is it composed of distinct modules?
- Are we naturally moral and good and only become evil through circumstance, or are we naturally evil and only made good through enforced circumstance?
- Why are men and women so different?

- Do men naturally want young and beautiful women—and as many as they can get?
- Do women naturally want rich and powerful men—and a bonded, monogamous, caring relationship?
- Are men naturally turned on (maybe too turned on) by the sight of a woman—or even a silhouette or cartoon of one?
- Do men like sex more than women do? If so, why?
- Just how much does a man's or woman's looks tell a member of the opposite sex about them and their value as a potential mate?
- Why do men get turned on by "lips like rubies, eyes like limpid pools, and skin like silk?" And why do women spend so much time and money trying to achieve and reinforce that appearance?
- Why do human males have such large penises relative to our nearest primate relatives the great apes?
- Do some human groups, on average, have larger (and therefore less ape-like) penises than other groups? If so, why?
- Why do human females have such large breasts relative to our nearest primate relatives the great apes?
- Do dominant Alpha Males have all the fun and leave the most descendants, or do "Sneaky Fuckers" beat them at their own game?
- Why do women have orgasms?
- Why do cute, lovable children so quickly transmogrify into wild, ungrateful teenagers?
- Why, as we grow old, do we feel, in the words of then retiring Supreme Court Justice, the late Thurgood Marshall, that we're "just fallin' apart?"
- Do men naturally form power pyramids and hierarchies while women naturally form cliques?
- Do we naturally partition the world into Us v. Them?
- Does maternal instinct explain why moms usually act like moms, while dads all too often act like cads?
- Do we naturally prefer those who physically resemble us and find them to be more like us in other ways as well?

Are such questions even scientifically meaningful or do they more properly fall in the realms of religion, literature, or politics? They are certainly great openers to liven up even the dullest party. But the new and emerging field of evolutionary psychology, building on work from Charles

Darwin's *The Descent of Man* and *The Expression of Emotion in Man and Animals*, tells us that the answers to these age-old questions, dear Brutus, are in our evolutionary history and our genes. And they claim they've got the "bloody daggers" to prove it!

This introduction cannot examine the evolutionary argument on each of these points. Instead, it merely outlines the case and describes the type of evidence and the nature of the arguments to be placed before you, the skeptical jury. The references in the bibliography provide a more complete "transcript."

FROM SURVIVAL OF THE FITTEST TO INCLUSIVE FITNESS

The fundamental theorem upon which evolutionary psychology is based is that behavior (just like anatomy and physiology) is in large part inherited and that every organism acts (consciously or not) to enhance its inclusive fitness—to increase the frequency and distribution of its selfish genes in future generations. And those genes exist not only in the individual but in his or her identical twin (100%), siblings (on average, 50%), cousins (on average, 12.5%) and so on down the kinship line. (Thus, aid to and feelings for relatives makes evolutionary sense.)

This revision and extension of Darwinian evolution, from "survival of the fittest" to inclusive fitness, was worked out primarily by George Williams (in the US) and by William Hamilton and John Maynard Smith (in the UK) in the 1960s, with some clever twists added by Robert Trivers (in the US) in the 1970s. How efficiently can the Darwinian mill grind they asked? It largely depends on the type of grain fed in. Darwinian selection operates most effectively if the units on which it is working:

1. are more, rather than less, variable;
2. have shorter, rather than longer, lifetimes;
3. are more heritable, rather than environmental.

Richard Alexander (1979) has argued convincingly that "genes are the most persistent of all living units, hence on all counts the most likely units of selection. One may say that genes evolved to survive by reproducing, and they have evolved to reproduce by creating and guiding the conduct and fate of all the units above them" (p.38, emphasis Alexander's).

Implicit in this reasoning is the conclusion that species and populations (races) are very unlikely units of selection. Hence, all talk of individuals doing things, especially dying, for the good of the species or the race, appear improbable if not downright impossible. But if that is the case, then how could any sort of cooperative behavior, of which there are as many examples all around us as there are of competitive behavior, have ever evolved?

7

Well, humans, like most complex species, don't pass on their genes by simply dividing and producing exact replicas of themselves the way amoebas do. It takes at least two, not only to tango, but to reproduce. While you need not share any genes with your mate, you must share some, but not necessarily all of them with your relatives (except in the interesting case of an identical twin, who shares all your genes). Work out the arithmetic and it produces some interesting consequences in terms of whom you should help and when, as summarized in Figure 1.1 (adapted from Alexander, 1979). Rather than anything so simple as either "every man for himself" or "all for one and one for all," Figure 1.1 shows that, like it or not, you're stuck in a complex, time-directed matrix of cooperation, competition, trust, and deception with all your blood relatives and even those you might think are blood relatives.

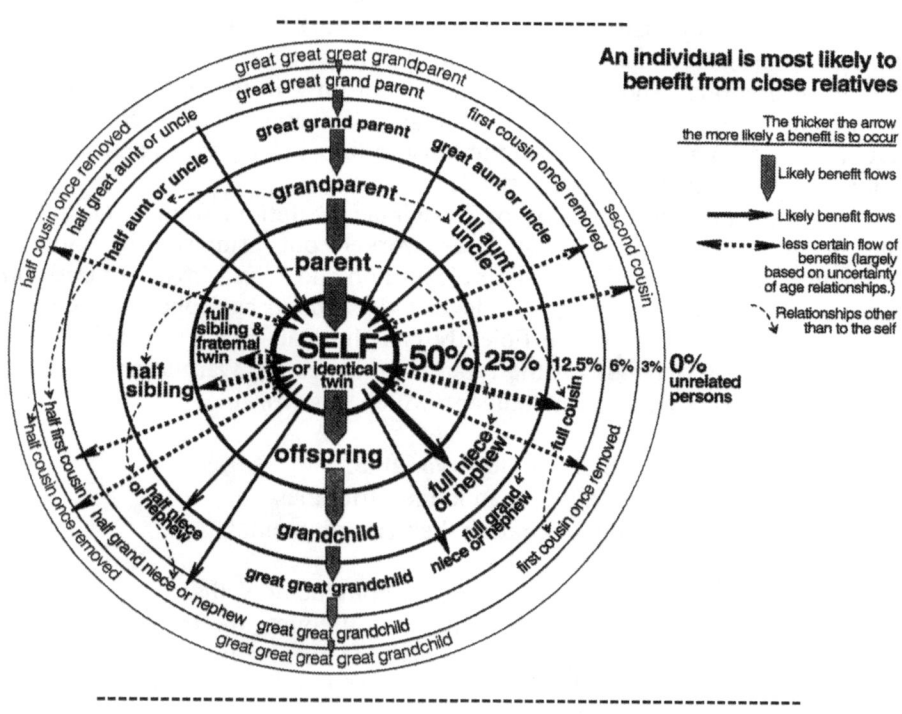

Figure 1.1: Proportionately thicker than water. According to evolutionary psychology theory, the percentage of genes (on average) shared by various degress of kinship should predict the amount of benefits received from a given individual. The right side of the diagram shows relatives resulting from monogamy; the left side, polygny. [As adapted in

Skeptic, and used with their permission, from Alexander, 1979; Figure 4, p.44]

Appropriately enough, you watch out for Number 1 first; your parents, children, and full siblings next; and so on in order of decreasing genetic similarity. But given that time's arrow flies in one direction only, you have a better chance of passing on your genes by helping your children than by helping your aging parents.

SYMONS SAYS

What does evolutionary theory predict you should expect from your mates? The answer is even more disconcerting. A corollary to the fundamental theorem is that the differences between males and females in humans, just as in most mammalian species, are readily explainable in terms of differential parental investment. That is, the male contributions to the reproductive process—lots of sperm and a few minutes of light work—are plentiful and cheap, short and pleasurable; while the female contributions— eggs and months of pregnancy—are rare and expensive, long, dangerous, and often painful. Given that, the best way for a male to maximize his inclusive fitness is to...well, diversify his genetic portfolio; while the best way for a female to insure the survival of the baby she has invested so much time and effort in is to try and get that guy to meet his monthly payments.

In *The Evolution of Human Sexuality* (p. 27, 1979), anthropologist Donald Symons provides evolutionary psychology's point-by-point reply to "the horny little devil's" soliloquy on men and women:

1. Intrasexual competition generally is much more intense among males than among females, and in preliterate societies competition over women probably is the single most important cause of violence.
2. Men incline to polygyny, whereas women are more malleable in this respect and, depending on the circumstances, may be equally satisfied in polygynous [one male—multiple females], monogamous, or polyandrous [one female—multiple males] marriages.
3. Almost universally, men experience sexual jealousy of their mates. Women are more malleable in this respect, but in certain circumstances, women's experience of sexual jealousy may be characteristic as intense as men's.

9

4. Men are much more likely to be sexually aroused by the sight of women and the female genitals than women are by the sight of men and the male genitals. Such arousal must be distinguished from arousal produced by the sight of, or the description of, an actual sexual encounter, since male-female differences in the latter may be minimal.

5. Physical characteristics, especially those that correlate with youth, are by far the most important determinants of women's sexual attractiveness. Physical characteristics are somewhat less important determinants of men's sexual attractiveness; political and economic prowess are more important; and youth is relatively unimportant.

6. Much more than women, men are predisposed to desire a variety of sex partners for the sake of variety.

7. Among all peoples, copulation is considered to be essentially a service or favor that women render to men, and not vice versa, regardless of which sex derives or is thought to derive greater pleasure from sexual intercourse.

To many, this sets a new standard in arguing for the inherent and therefore inescapable nature of the double standard. What evidence is there to support the argument that male-female differences are so deeply rooted in our nature? Anthropologists Lionel Tiger and Robin Fox argued in 1971 in *The Imperial Animal* (see Chapter 3) that if "we look at enough primates to see what we all have in common, we'll get some idea of what it was we evolved from. If we see what we had to change from to get to be what we are now, it might help to explain what we in fact are."

OF BELLES AND BALLS

Figures 1.2 and 1.3 are adapted from Jared Diamond's *The Third Chimpanzee* (pp. 73-74). They compare the relevant male and female anatomy for humans and our nearest living relatives, the great apes.

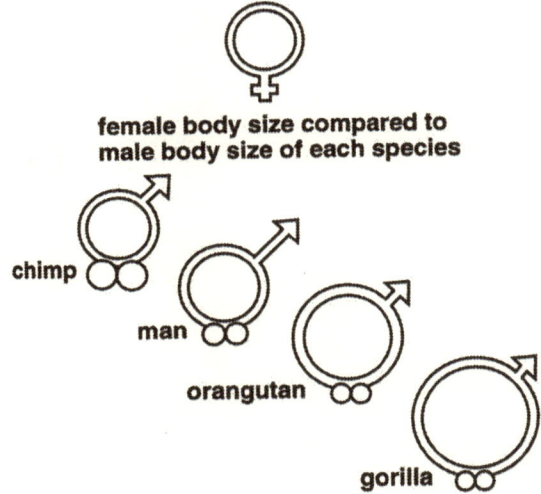

Figure 1.2: A species of welll-hung males…The relative body size, penis length, and testicle size for humans compared to the great apes. Female body size, represented by the top circle is compared t the male body size of each species. The smaller twin "balls" represent testes weight as a proportion of total body weight. Arrows represent the relative legnth of the erect penis, across species. [As adapted in *Skeptic*, and used with their permission, from Diamond, 1992; Figure 4, p. 73]

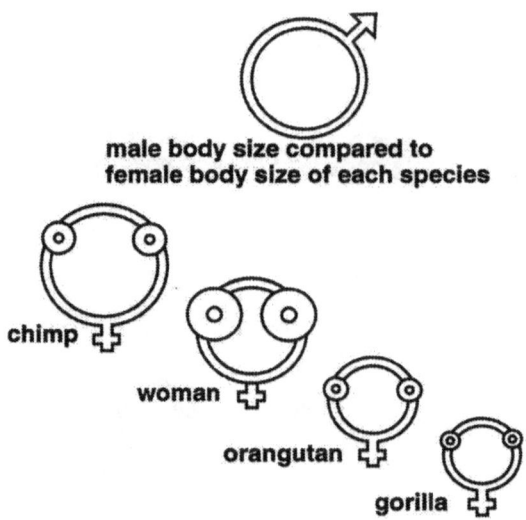

male body size compared to
female body size of each species

chimp

woman

orangutan

gorilla

Figure 1.3:...And stacked females. The relative body size and breast size for humans compared to that of the great apes. Male body size, represented by the top circle is compared to the female body size of each species. Relative breast size to body size for the female of each species is represented by the smaller "sets" of circles. [As adapted in *Skeptic*, and used with their permission, from Diamond, 1992; Figure 5, p. 74]

First look at the amount of sexual dimorphism in the four species. As Diamond notes, "chimps of both sexes weigh about the same; men are slightly larger than women, but male orangutans and gorillas are much bigger than females" (p. 73). These are interesting facts from comparative anatomy, but what do they have to do with behavior? Throughout the animal kingdom, polygynous species (i.e., those in which each dominant male breeds with multiple females), are sexually dimorphic. This makes sense from an evolutionary point of view. The only way a male can pass on his genes is to breed with a female, and to better the odds, the more the merrier. But since there are only so many females to go around, from day 1 males are in competition with other males for those females. An arms race begins in which males are selected for their ability to win out against other males for access to the females. And since nothing escalates like an arms race, you end up with male gorillas and orangs that are not only twice the size of the females, but armed with huge canines, and loaded with secondary sexual characteristics like crested heads and silver backs that are easily recognizable at a distance and help to attract mates.

Chimps, on the other hand, show little sexual dimorphism, less even than humans. The gibbon (an ape, but not a great one) shows the least sexual dimorphism. Males and females look identical at a distance and the gibbons' strict adherence to monogamy should win an award from the Moral Majority (though that would mean acknowledging man's common primate ancestry and therefore ditching creationism). Going simply by the dope sheet of sexual dimorphism, an evolutionary handicapper would bet the rent that *Homo sapiens* would, by nature, be mildly polygynous. And he'd walk away from the pay window a big winner. A cross-cultural analysis of 853 societies revealed that 83% of them are polygynous. Polygyny occurs frequently, even when legally prohibited. There are an estimated 25,000 to 35,000 polygynous marriages in the US; a study of 437 financially successful American men found that some maintained two separate families, each unknown to the other (Buss, pp. 177-178). Polyandry (one female with multiple males), on the other hand, is "virtually absent" among hunter/gatherers and confined to "agriculturalists and pastoralists living under very difficult economic conditions" and disappears quickly "when more usual conditions are present" (Symons, p. 225).

To move on from gross anatomy to gross discourse, if the male gorilla is so big and tough, how come he has such small balls? How does evolutionary theory account for those differences in testicle, penis, and breast size? It may be a tough climb to the top of the male gorilla dominance pyramid, but once there, things become quieter. Until dethroned, you have virtually uncontested access to all the females, so sex is no big thing. In fact, the dominant male with a harem of females "experiences sex as a rare treat: if he is lucky, a few times a year" (Diamond, p. 73). So just a little bit of sperm goes a long way to insuring the male gorilla's inclusive fitness.

For the minimally sexually dimorphic chimp, things get a little dicier. Chimps do have power pyramids. Compared to the gorilla and the orang, their hierarchies are so complex that Frans de Waal entitled his study of them *Chimpanzee Politics*. Getting to the top and staying there calls more for the skills of a Machiavelli than of a Mike Tyson. Dominant males have frequent though not exclusive access to the females. Rather than simply their bodies, it is their sperm that must compete against those of their fellow dominants, as well as those of the occasional "sneaky fucker." And all of this follows directly from one of the triumphs of evolutionary biology – The Theory of Testicle Size – to wit, "species that copulate more often need bigger testes; and promiscuous species in which several males routinely copulate in quick sequence with one female need especially big testes (because the male that injects the most semen has the best chance of being the one to fertilize the egg). When fertilization is a competitive lottery, large testes enable a male to enter more sperm in the lottery" (Diamond, p. 72).

Humans, according to evolutionary theory, should therefore be intermediate between chimps and gorillas both in polygyny and in promiscuity—and the data fit the prediction. I leave it to the reader to speculate as to what the evolutionary result would be if groups of religious cultists (in which the leader tries to monopolize the females) and outlaw biker gangs (who after all gave us the term "gang bang") were to each pursue their own evolutionary path, separate from the rest of human society.

Diamond provides more hard anatomical data (p. 75):

The length of the erect penis averages 1-1/4 inches in a gorilla, 1-1/2 inches in an orangutan, 3 inches in a chimp, and 5 inches in a man. Visual conspicuousness varies in the same sequence: a gorilla's penis is inconspicuous even when erect because of its black color, while the chimp's pink erect penis stands out against the bare white skin behind it. The flaccid penis is not even visible in apes.

To date, however, there is no adequate evolutionary explanation of the between-species differences in penis size. J. P. Rushton has offered a very controversial explanation of the mean differences in penis size between various racial groups within the human species. His letter to *Skeptic* (Vol. 3, No. 4, pp. 22-25), with an accompanying table, summarizes his argument that there is a "tradeoff" between cognitive assets (brain size and IQ score) and reproductive assets (penis size and gamete production). Both neurons and gametes are expensive and Rushton's data are replicable, but most evolutionary biologists and psychologists do not accept his interpretation.

Rushton's work highlights two important differences among evolutionary explanations of behavior. Evolutionary explanations of genetic differences between individuals, and especially between groups of individuals, have an air of an earlier Social Darwinism which many today find downright offensive. Which is not to say that they are, for that reason, factually wrong. But most of today's evolutionary psychologists are concerned with the universals of human nature, not the differences. They argue that "genetic differences among individuals surely play a role, but perhaps a larger role is played by genetic commonalities: by a generic, species-wide developmental program that absorbs information from the social environment and adjusts the maturing mind accordingly." They therefore believe that "future progress in grasping the importance of environment will probably come from thinking about genes" (Wright, p. 9).

And whereas Rushton and others, located on the right or pro side of *The Bell Curve* controversy, argue for a unitary view of the mind (usually manifested in a single trait variously referred to as intelligence, IQ, cognitive ability, or psychometric *g*) on which all individuals (and even groups) can be measured and ranked from top to bottom ("alphabetically by height" as legendary New York Yankee manager Casey Stengel once put it), most of today's evolutionary psychologists argue that evolution would rather select for distinct mental modules. In their view, evolution can give males a "love of offspring" module, and make that module sensitive to the likelihood that the offspring in question is indeed the man's. But the adaptation cannot be foolproof. Natural selection can give women an "attracted to muscles" module, or an "attracted to status" module, and...it can make the strength of those attractions depend on all kinds of germane factors.... As Tooby and Cosmides say, human beings aren't general purpose "fitness maximizers." They are "adaptation executors." The adaptations may or may not bring good results in any given case, and success is especially spotty in environments other than a small hunter-gatherer village (Wright, pp. 106-107).

In the view of most evolutionary psychologists, the modules may differ in effectiveness from one individual to another, but given the number of different modules, their effect is to "average out" individual differences to the point where any attempt to "line everyone up" on a single dimension is as nebulous as Casey's syntax.

Now let's look at the females. "Human females are unique in their breasts, which are considerably larger than those of apes even before the first pregnancy" (Diamond, p. 74). Since the female gorilla and her baby are comparable in size to their human counterparts, the bulk of the huge (by primate standards) human female breast consists of fat, not milk glands, and breast size varies greatly among human females without affecting their ability to nurse young. Thus, the explanation cannot be based on the need to nurse infants. Rather, human female breasts are secondary sexual characteristics that evolved to attract mates. According to Desmond Morris (1967), this took place along with the switch from front-to-rear to front-to-front mating, the pendulous shape and cleavage of the breasts mimicking the pre-existing attractiveness of the female buttocks. This also, according to the theory, explains why men find other pendulous shapes (like ear lobes) and other cleavages (like toes in low-vamped shoes) such a turn-on.

And while we're on the subject, what other female attributes turn men on? Gentlemen prefer young, nubile women, with lips like rubies, eyes like limpid pools, skin like silk, breasts like a milch cow, and legs like a race horse. According to evolutionary theory, this is not the result of either Hollywood or Madison Avenue, but because all of these features have served as cues to a female's health, reproductive potential and sexual availability over the course of human evolutionary history. Evolution has built into every red-blooded male a desire to find "Pornotopia"—the fantasy land where "sex is sheer lust and physical gratification, devoid of more tender feelings and encumbering relationships, in which women are always aroused, or at least easily arousable, and ultimately are always willing"

(Symons, p. 171). The entire cosmetics, fashion, and pornography industries are attempts to create Pornotopia here on Earth.

Figure 1.4, adapted from Daly and Wilson (1988) depicts human female reproductive value, calculated in terms of expected live births among hunter/gatherers, as a function of female age. This curve parallels the curve for men's preferences in females as determined in cross-cultural studies (Buss, pp. 49-60; Symons, pp. 187-200).

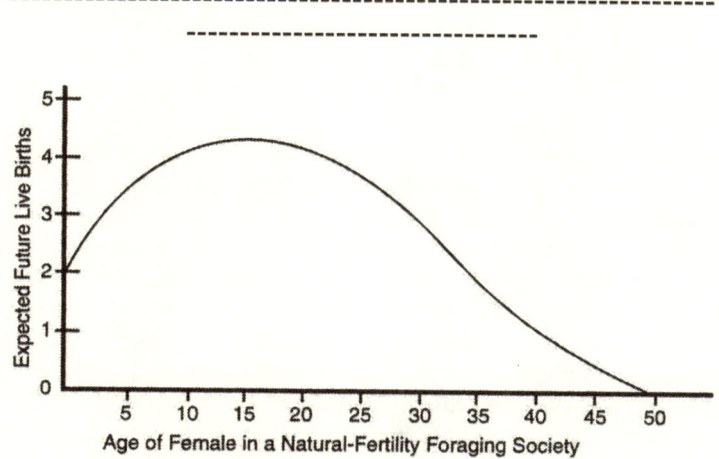

Figure 1.4: Babes as baby factories. This figure shows the relative likelihood of a female bearing children in hunter/gatherer societies where surviving to maturity is far less certain than in modern industrial societies; thus the average 12 year old girl is likely to bear more children over the remainder of her lifetime than the average female infant (who may not reach 12) or the average 30-year old (whose remaining childbearing years are limited). Across societies, males' desire for females traces a similar curve. [As adapted in *Skeptic*, and used with their permission, from Daly and Wilson, 1988; Figure 4.3, p. 74]

Men naturally prefer young women because they provide the most reproductive potential for passing on the male's genes. If anything, males are biased toward selecting females before reproductive age in order to insure that no other male has beaten them to the finish line. From an evolutionary perspective, the least wise thing a male can do is to divert his hard-earned resources to rearing another man's child. Indeed, evolutionary psychologists would argue that this is why cuckolds are universally held in such low regard.

MURDER 1, INCEST 0

According to evolutionary theory, sex is a service women provide to men in return for resources. Evolutionary psychologists Martin Daly and Margo Wilson note that (p. 188, emphasis theirs):

> marriage is a contract not between husband and wife, but between *men*, a formalized transfer of a woman as a commodity. And indeed when one examines the material and labor exchanges that surround marriage, it does begin to look like a trafficking in women. In our society, as in many, a father *gives* his daughter in marriage. Men purchase wives in the majority of human societies, and they often demand a refund if the bargain proves disappointing. Although the relatively rare practice of dowry might be construed to mean that who pays whom is arbitrary and reversible, dowry and bride-price are not in fact opposites: A bride-price is given as compensation to the bride's kin, whereas a dowry typically remains with the newlyweds.

Figure 1.5 (adapted from Daly & Wilson, p. 189) summarizes the exchange considerations at marriage in a cross-cultural comparison of 860 societies and emphasizes the universality of compensation for rights to female reproductive capacity.

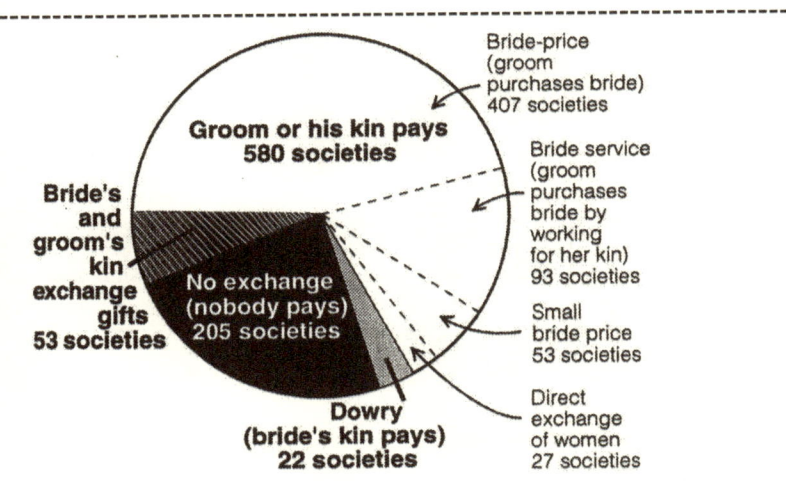

Figure 1.5: Who pays to play. In most human societies, the male (and his family) pay the female (and her family) for "leasing rights" to her "depreciating assets" (i.e., marriage). [As adapted in *Skeptic*, and used with their permission, from Daly and Wilson, 1988; Figure 9.1, p. 189]

Even worse from the point of view of the male and his family than failure by the female to live up to her part of the contract is the thought that the male's investment in resources may be going into a competitor's product. Figures 1.6 and 1.7 (adapted from *Homicide* by Daly and Wilson) show that child abuse and even murder are much more common for adoptive parents than for natural parents.

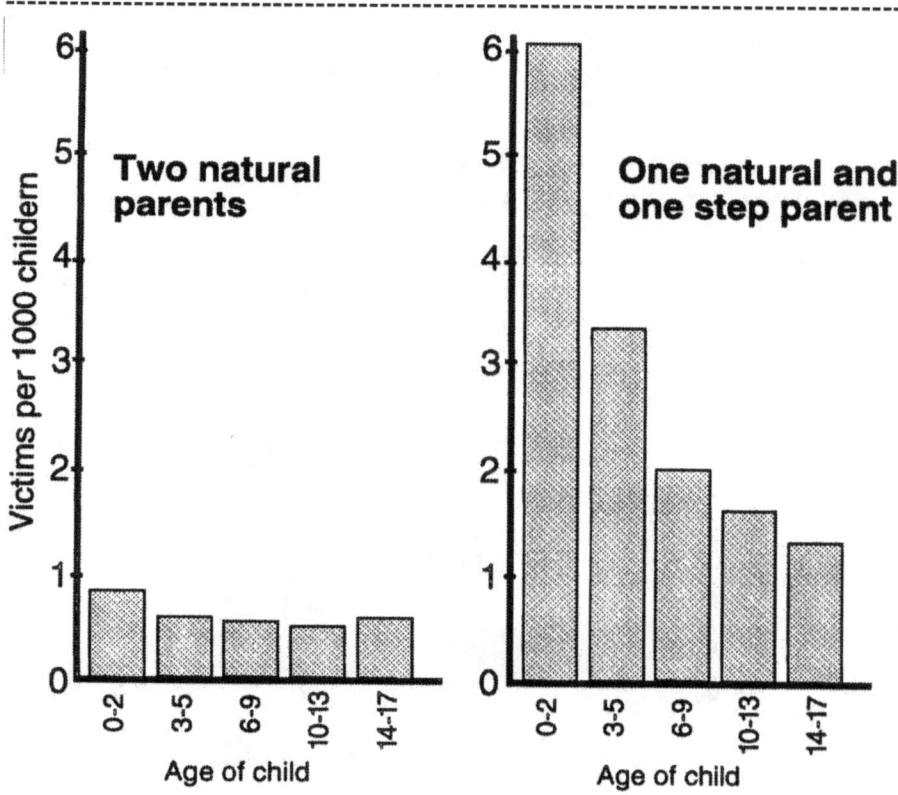

Figure 1.6: Who abuses children? There is a much greater per capita rate of validated child abuse cases by adoptive parents than there is by natural parents. [As adapted in *Skeptic*, and used with their permission, from Daly and Wilson, 1988; Figure 4.7, p. 86]

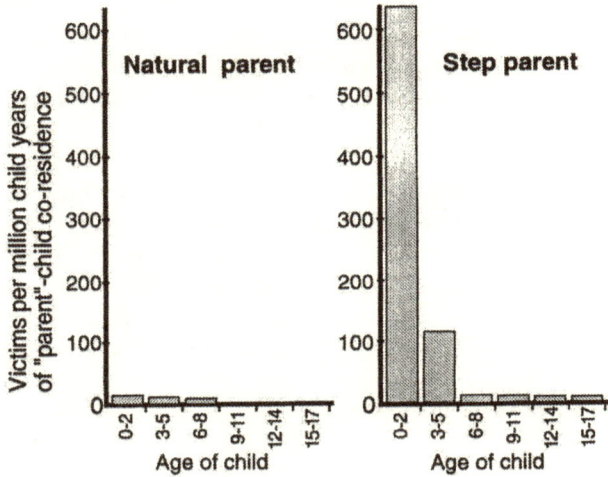

Figure 1.7: Cinderella was lucky. There is a much greater risk of being killed by an adotive a parent tghan of being killed by a natural parent. [As adapted in *Skeptic*, and used with their permission, from Daly and Wilson, 1988; Figure 4.9, p. 90]

While evolutionary theory predicts a certain level of parent-child and sibling rivalry, its predictions are contrary to another mainstay of social science—the Freudian Oedipus Complex. Under evolutionary theory, fathers have a strong vested interest in their son's well-being; provided, of course, it is their son. As sons mature, they may in fact compete with their fathers for status and for females (as daughters may compete with their mothers for males), but not for their *own* mother (or father). Many evolutionists argue that, given the decreased viability of children born out of incest, selection has created an incest taboo, especially against mother-son incest. The comparative ethnographic data support the existence of the incest taboo, not the Oedipus complex (Alexander, p.165; Wright, pp. 315-316).

THEY SAY THAT BREAKING UP IS HARD TO DO
FISHER'S DIVORCE LAW SAYS IT ISN'T

Evolutionary psychology provides explanations not only of why we pair up, but why we split up. Conservative social critics have decried the alarming increase in divorce in the US since the 1960s, and variously attribute it to removing Bible reading from the public schools, rock 'n' roll, TV and movies, liberal social welfare programs, decriminalization of abortion, women's lib, and even the teaching of evolution. The evolutionary perspective, on the other hand, leads one to see lifetime monogamy as the exceptional result of an increased level of social pressure rather than as the rule for humans.

Anthropologist Helen Fisher has gathered divorce data from 62 societies around the world (Figures 8 and 9). She finds that "human beings in a variety of societies tend to divorce between the second and fourth years of marriage, with a divorce peak during the fourth year" (p. 360). She also finds that the divorce statistics for the US in 1986, well past the sexual revolution of the 1960s, fit the same pattern, with most divorces taking place between the second and third year of marriage (p. 362).

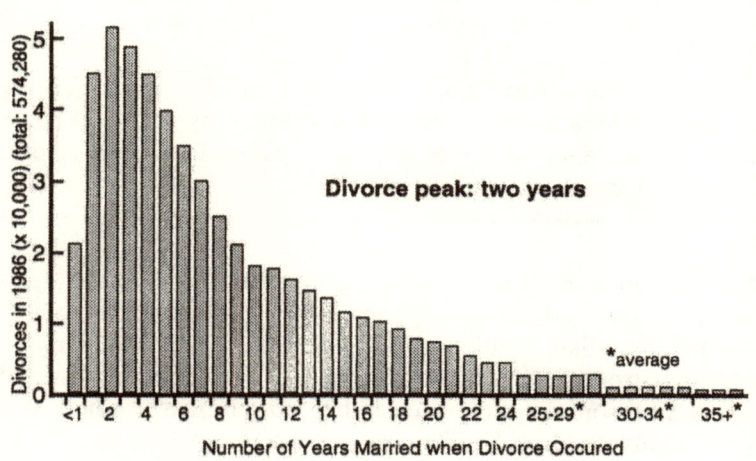

Figure 1.8: The less than seven year itch. The consistency of divorce pattern for 62 different societies suggests an evolutionary mechanism based on the typical period of infant dependency. [As adapted in *Skeptic*, and used with their permission, from Fisher, 1992; Figure 4, p. 362]

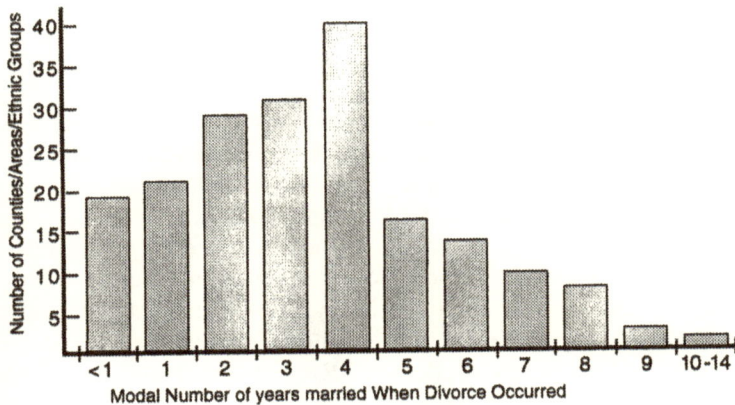

Figure 1.9: Divorce American style ain't so different. The divorce pattern for America in the midst of sexual revolution parallels that for other societies. [As adapted in *Skeptic*, and used with their permission, from Fisher, 1992; Figure 2, p. 360]

Fisher's evolutionary explanation attributes the universality of the divorce statistics to the "remarkable correlation between the length of human infancy in traditional societies, about four years, and the length of many marriages, about four years. Among the traditional! Kung, mothers hold their infants near their skin, breast-feed regularly through the day and night, nurse on demand, and offer their breasts as pacifiers. As a result of this constant body contact and nipple stimulation, as well as high levels of exercise and a low-fat diet, ovulation is suppressed and the ability to become pregnant is postponed for about three years" (p.153). She therefore concludes (p. 154):

> The modern divorce peak—about four years—conforms to the traditional period between human successive births—four years....Like pair-bonding in foxes, robins, and many other species that mate only through a breeding season, human pair–bonds originally evolved to last only long enough to raise a single dependent child through infancy, the first four years, unless a second child was conceived.

HUMAN, ALL TOO HUMAN

It may seem that either evolutionary psychology or the examples selected for this quick and dirty summary are more suited to tabloid TV than to *Skeptic* magazine. Were we trying to increase circulation by slumming to the lowest common denominator of human behavior? Well, evolutionary psychology has an answer for that one too. It is precisely because of our evolutionary history and the importance of maximizing inclusive fitness that humans in all cultures, throughout history, have found such lurid tales so irresistible. Some may prefer them told with British accents on *Masterpiece Theatre*, rather than in the dialect of Rap or the twang of Country & Western, but the archetypal themes are the same and evolutionary psychology tells us that they will never go away.

But just how scientific are these attempts to explain human behavior in evolutionary terms? To what extent do the questions we ask automatically set up the answers we get? After all, as Cassius taunted Brutus, we are sometimes masters of our own fate! To what extent are human nature and individual and group differences scientifically meaningful concepts, rather than the social constructions of learning and experience, political and economic conditions? Is there any scientific there there?

So here then, ladies and gentlemen of the jury, is the issue at hand: Should we accept as a default hypothesis that human behavior, and the similarities and differences in behavior between individuals and groups, are the result of a complex interaction of the genes that reflect our evolutionary history as well as the environment in which we find ourselves? Or should we opt for the statistically null hypothesis that any invocation of genes and evolution to explain human behavior must be proved beyond a reasonable doubt? If nothing else, when you finish reading Part I - The Evolutionary Psychology Debates, I think you will be forced to consider the words of Nobel Prize Winner and co-discoverer of DNA James Watson, "Charles Darwin will eventually be seen as a far more influential figure in the history of human thought than either Jesus Christ or Mohammed."

References

Alexander, Richard. 1979. *Darwinism and Human Affairs.* Seattle, WA: University of Washington Press.

Buss, David. 1994. *The Evolution of Desire.* New York: Basic Books.

Daly, Martin and Wilson, Margo. 1988. *Homicide.* New York: Aldine.

de Waal, Frans. 1982. *Chimpanzee Politics: Power and Sex Among the Apes.* Baltimore: Johns Hopkins.

Diamond, Jared. 1992. *The Third Chimpanzee.* New York: Harper.

Fisher, Helen. 1992. T*he Anatomy of Love: The Natural History of Monogamy, Adultery, and Divorce.* New York: Norton.

Morris, Desmond, 1967. *The Naked Ape.* New York: McGraw Hill.

Rushton, J. P. 1995 *Race, Evolution, and Behavior.* New Brunswick, NJ: Transaction.

Symons, Donald. 1979. *The Evolution of Human Sexuality.* Oxford University Press.

Tooby, John, and Cosmides, Leda, "The Psychological Foundations of Culture" in Barkow, Jerome; Cosmides, Leda; and Tooby, John. 1992. *The Adapted Mind: Evolutionary Psychology and the Generation of Culture.* New York: Oxford.

Wright, Robert. 1994. The *Moral Animal: The New Science of Evolutionary Psychology.* NY: Vintage.

Frank Miele

CHAPTER 2

DARWIN'S DANGEROUS DISCIPLE

AN INTERVIEW WITH RICHARD DAWKINS

Richard Dawkins, fellow of New College, Oxford, is one of the leading thinkers in modern evolutionary biology. He is also one of the best writers on the subject. His books *The Selfish Gene* (1976; expanded, 2nd edition, 1989), *The Extended Phenotype* (1982, 1989), *The Blind Watchmaker* (1986), and *River Out of Eden* in the Brockman Science Masters Series (Basic Books, 1995), have introduced the terms "Blind Watchmaker," "Selfish Genes," "Memes," "Green Beards," "Biomorphs," "Arms Races, Sheriff Genes, and Outlaw Genes," and "Mind Viruses" to both professional and popular audiences. *The New York Times* has described his books as "the sort of popular science writing that makes the reader feel like a genius." While in California this summer, Dawkins spoke before the Human Behavior and Evolution Society on the "Evolution of Perceptual Models," and at the Skeptics Society on the "Fallacies of Creationism." He also took time to speak with *Skeptic* magazine on the triumphs, limitations, uses, and abuses of Darwinism. Never one to shy away from controversy, here is what the man *Wired* magazine termed "the Bad Boy of Evolution" had to say about Darwinism, extra-terrestrial life, religion as a virus of the mind, morality, politics, punctuated equilibrium, and the future of evolutionary biology.

Figure 2.0: Richard Dawkins

Miele: In your latest book, *River Out of Eden*, which is a best seller in the UK, you use deep-sea bacteria that metabolize sulfur (rather than oxygen) to illustrate how evolution takes place in a series of successive steps. Does the existence here on Earth of an alternative metabolic "fuel," in some sense make it more probable that there could be life elsewhere in the universe, perhaps using a different base than carbon?

Dawkins: That's surely got to be right, hasn't it? You can speculate in a science fiction way about alternative biochemistries for life, but if you couldn't find anything on Earth moving ever so slightly towards an alternative biochemistry, that would argue against the idea. But when you do find an alternative biochemistry for life here on Earth, that makes it more plausible that somewhere else in the universe there's got to be an alternative form of life.

Miele: What then is the *sine qua non* of life? What raw materials and conditions are necessary for life to exist?

Dawkins: Well, you need raw materials that can self-replicate. I would have to be more of a chemist than I am to know how likely it is that you are going to get such molecules. I should very much like to direct chemists toward devising an alternative hypothetical chemistry that supports self-replication, a whole alternative system that could, in principle, give rise to life. The fundamental principle that will be required is self-replication. Chemists have begun to look at auto-catalytic functions in chemistry where at least some of the prerequisites are present. The *sine qua non*, as you say, is self-replication. I don't know how difficult it would be to achieve that chemically.

Miele: How likely do you think it is that "intelligent" life exists somewhere else in the universe?

Dawkins: At first glance, one might think that the really difficult step is getting life at all. Then once natural selection has gotten going (since the origin of life is really the origin of natural selection), you can proceed by an orderly progressive sequence through the evolution of some kind of information processing apparatus on to intelligence. On the other hand, if you look at what's actually happened on this planet, it probably took less than a billion years from the origin of the planet, under fairly unfavorable initial conditions, to produce life. But intelligence of a high order has only come about in the last couple of million years, perhaps. So it does seem that on this planet at least there has been a rather short interval from the origin of the planet to the origin of life and then a very, very long interval between the origin of life and the origin of intelligence.

Miele: Are you then saying that the origin of intelligence is the bigger step?

Dawkins: It's not my inclination to say that, but this disparity in time scale is the only data we have. We only have one sample—life on this planet. But for that fact, my personal inclination would have been to suggest that the origin of intelligence is not that difficult once you've got life. I'm quite intrigued by the thought that maybe it's the origin of life that's not that difficult.

Miele: If so, what are the defining qualities of such "intelligence?" I'm thinking here of concepts such as Immanuel Kant's list of a priori ideas—time and space, number, cause and effect. Could, for example, a form of life evolve in whose mental map time's arrow went backwards, or in no specific direction?

Dawkins: As for what one means by intelligence, I haven't really thought about that. You posed the hypothetical question, "Could there be a life form whose concept of time goes backward?" I can't imagine what that would look like. But I haven't thought about it enough.

Miele: Of other species we know about here on Earth, we're very familiar with cats and dogs. I was always amazed (and delighted) how my dogs could coexist with me given that their set of sensory inputs was so different from mine—they see in black and white, not in 3D; they use scent and vision in approximately opposite proportions. Yet they can come up with an equivalent map of the world so that they can fetch the paper, protect us from intruders, comfort us in distress. What does this tell us about the evolutionary process and how it molds not only the bodies but the cognitive maps of different species?

Dawkins: That's an intriguing point. One thing you could say about dogs is that they have been domesticated, and a good deal of the domestication has been an inadvertent selection for coexisting with humans. While their wild ancestors, the wolves, do have facial expressions and other gestures which they use to communicate with each other, it's probable that domestic dogs have been selected to have more human feelings and facial expressions. So while dogs don't smile, they do other things with their eyes that appeal to humans. Maybe they have been shaped to be a bit less wolf-like and a bit more human-like, not by deliberate artificial selection, but by artificial selection nonetheless.

Miele: In the 1930s, von Uexküll used the term *Umwelt* to describe the different "real worlds" that animals construct based upon their differing sensory systems. He even built mechanical devices to try and create their perceptual *Weltanschauungen.* He manufactured optical devices to simulate the compound eyes of insects to allow one to see "what they saw." With virtual reality now a reality, we could certainly do such things at a more sophisticated level than von Uexküll did. Do you think this might be a potentially valuable line of research?

Dawkins: Von Uexküll used the concept of the *Umwelt* to explore the differences between the perceptual worlds of different animals. He tried to find a way to "think himself into" the *Umwelt* (the perceptual world) of a bee or a bat, for example, by seeing the polarization of light or by seeing into the ultraviolet range of the spectrum and thus probably not seeing images as we see images at all. I think it's a very important thing to do that, partly as a metaphor for "getting outside yourself" and seeing another point of view. We have an immensely human-centered view of things such as ethics and morality. Even if we pay lip service to being evolutionists, many people still think according to the Judeo-Christian view that all things have been put on Earth for the benefit of humanity and that the only justification for scientific research is if it benefits humanity. I think it's a salutary lesson to try to "think yourself into" the *Umwelt* of another species. But as I said in my talk last night before the Human Behavior and Evolution Society, I suspect that the perceptual world of other species possibly may not be as different from our own as you might think, even though they get their information through different physical media.

Miele: So then doesn't natural selection force us to assume that time's arrow flies in a certain direction? Doesn't natural selection force us to operate on the basis of Kant's a priori ideas and Piaget's operations?

Dawkins: Yes, I agree with that.

Miele: In your speech the other night you said that the perceptual systems of animals represent the world as their near, or possibly even far, ancestors constructed them based upon natural selection. Can the world evolve faster than the sensory systems of the animals? Are many animals living today in a sensory world that no longer exists, as when the moth flies into the candle flame?

Dawkins: When a moth flies into a candle flame presumably it is responding to the candle flame as if were a celestial object at optical infinity and acting appropriately to that situation, not the one it is in fact currently facing. It frequently happens that the real world evolves faster than an animal's cognitive map of it.

Miele: Does that ever happen to human beings?

Dawkins: Human beings are completely surrounded by the equivalent of "candle flames." Notorious examples are our desire for sugar and fat—in nature the rule is, whenever you can get them, eat them. But when there's a surplus of those substance, they become bad for you. Most of what we strive for in our modern life uses the apparatus of goal seeking that was originally set up to seek goals in the state of nature. But now the goal-seeking apparatus has been switched to different goals, like making money or hedonistic pleasures of one sort or another. Natural selection equips us with "Rules of Thumb," which in a state of nature have the effect of promoting the survival of our selfish genes. The Rules of Thumb go on, even though in this world of "candle flames" they no longer promote our inclusive fitness.

Miele: Would you consider increasing population and war to be examples of candle flames?

Dawkins: Increasing population itself is not an individual behavior pattern. It's a consequence of many things which are manifestations of individual behavior in a collective environment. To take a much simpler case, the dominance hierarchy is a manifestation of attacking and subservience between pairs of individuals, but the dominance hierarchy itself is not something that natural selection favors or disfavors. What natural selection favors or disfavors is the individual behavior of which the dominance hierarchy is a manifestation. I would put war and overpopulation in that category.

Miele: In *River Out of Eden*, you also say that, "Science shares with religion the claim that it answers deep questions about origins, the nature of life, and the cosmos. But there the resemblance ends. Scientific beliefs are supported by evidence, and they get results. Myths and faiths are not and do not" (p. 33). But doesn't one first have to make the choice or decision to use pragmatism as the standard by which we judge? That is, we must first agree to base our decisions on what works, rather than on revelation or intuition. Isn't the most we can ask of the religious crowd, "Either lay hands on flat tires and pray for the sick, rather than taking them to a mechanic or a doctor, or if you are not willing to be consistent, just shut up and go away?" Doesn't the religious view amount to, "When we're afraid, we seek God. When God doesn't answer our prayers, blame it on the Devil?"

Dawkins: Yes, it's a kind of pathetic, childish response to some failure.

Miele: Then could you say that the reason that evolutionism is resisted so strongly is that our minds evolved to think in terms of personalities and entities rather than in terms of processes?

Dawkins: Yes, it's the idea that somebody has got to be responsible. It's what children do—the petulant throwing the tennis racquet on the ground, blaming it for their bad shot. So is the reflex to sue somebody when you slip on the ice and sprain your ankle. Again, somebody has got to be blamed. It doesn't occur to many people that nobody's to blame, there's just ice and it's slippery and you fell down.

Miele: Along that line, you have the following passage in *River Out of Eden*:

> In a universe of blind physical forces and genetic replication, some people are going to get hurt, other people are going to get lucky, and you won't find any rhyme or reason in it, nor any justice. The universe we observe has precisely the properties we should expect if there is, at bottom, no design, no purpose, no evil and no good, nothing but blind, pitiless indifference (p.133).

This sounds rather like physicist Steven Weinberg's, "the more the universe seems comprehensible, the more it seems pointless" (*The First Three Minutes*), or William Shakespeare's "a tale told by an idiot, filled with sound and fury, signifying nothing." Is that in fact your position?

Dawkins: Yes, at a sort of cosmic level, it is. But what I want to guard against is people therefore getting nihilistic in their personal lives. I don't see any reason for that at all. You can have a very happy and fulfilled personal life even if you think that the universe at large is a tale told by an idiot. You can still set up goals and have a very worthwhile life and not be nihilistic about it at a personal level.

Miele: Well, if we don't accept religion as a reasonable guide to "what is" or even a reasonable guide to "what ought to be," does evolution give us such a guide? Can we turn to evolution to answer not what is, but what ought to be?

Dawkins: I'd rather not do that. I think Julian Huxley was the last person who attempted to. In my opinion, a society run along "evolutionary" lines would not be a very nice society in which to live. But further, there's no logical reason why we should try to derive our normative standards from evolution. It's perfectly consistent to say this is the way it is—natural selection is out there and it is a very unpleasant process. Nature is red in tooth and claw. But I don't want to live in that kind of world. I want to change the world in which I live in such a way that natural selection no longer applies.

Miele: But given the clay from which we are made, doesn't natural selection make it relatively unlikely that some things will work? Doesn't Darwinism undercut the great socialist hope, "Why, because we will it so!"?

31

Dawkins: Some goals may be unrealistic. But that doesn't mean that we should turn around the other way and say therefore we should strive to make a Darwinian millennium come true.

Miele: But then isn't what we ought to do (as David Hume argued long ago) just a matter of preference and choice, custom and habit?

Dawkins: I think that's very likely true. But I don't think that having conceded that point, I as an individual should then be asked to abandon my own ethical system or goals. I as an individual can adopt idealistic or socialistic or unrealistic or whatever sort of norms of charity and good will towards other people. They may be doomed if you take a strong Darwinian line on human nature, but it's not obvious to me that they are.

Miele: I think that you are saying that many of the lessons of evolutionary biology about morality or ethics are contrary to what we might normally call morality or ethics in ordinary discourse. When we look back at the Old Testament and the New Testament, it seems there's a lot about how to maximize one's inclusive fitness. If that's the case, do such religious views have anything to tell us?

Dawkins: If it is true that some of the morality of the Old Testament, say, maximizes somebody's inclusive fitness, I don't think that has anything to tell us about what we ought to do.

Miele: Then wouldn't we be better to throw out all this half-baked religious mumbo jumbo and move on to something else?

Dawkins: Well yes, but that's obvious!

Miele: Do you think that group fitness is a meaningful concept in evolutionary biology? If so, does it play a role in discussions of evolution and morality?

Dawkins: I think it's logically meaningful, but I don't think it plays a role in evolution in the wild and therefore it doesn't play a role in anything else. I think it's of no importance.

Miele: Can we use Darwinism and natural selection to analyze other events in history? To put it in its crudest form, if Hitler had won W.W. II would that have proved that his system was better (in a Darwinian sense) than that of the Allies? Or does the fact that the Soviet Bloc crumbled tell us anything about the relative fitness of market-based economies versus command economies. If might (or at least survival and reproduction) doesn't make right (as well as everything else), what does?

Dawkins: I think it is not helpful to apply Darwinian language too widely. Conquest of nation by nation is too distant for Darwinian explanations to be helpful. Darwinism is the differential survival of self-replicating genes in a gene pool, usually as manifested by individual behavior, morphology, and phenotypes. Group selection of any kind is not Darwinism as Darwin understood it nor as I understand it. There is a very vague analogy between group selection and conquest of a nation by another nation, but I don't think it's a very helpful analogy. So I would prefer not to invoke Darwinian language for that kind of historical interpretation.

Miele: The publication of the controversial book, *The Bell Curve*, has set off yet another round in the seemingly never-ending nature-nurture (or heredity-environment) controversy. Is this in fact a false dichotomy? You had this passage in *The Selfish Gene* (1989, p. 37):

> No one factor, genetic, or environmental, can be considered as the single 'cause' of any part of a baby. All parts of a baby have a near infinite number of antecedent causes. But a *difference* [original emphasis] between one baby and another, for example a difference in length of leg, might easily be traced to one or a few simple antecedent differences, either in environment or in genes. It is *differences* that matter in the competitive struggle to survive; and it is genetically-controlled differences that matter in evolution.

Doesn't a passage like this indicate why E. O. Wilson's *Sociobiology*, *The Bell Curve*, and your own writings automatically set off a vehement, moralistic, and often hysterical reaction from a cadre of left-wing scientists?

Dawkins: On the face of it, I don't know why you lump those three books together. When I was talking about genetic differences as being analyzable, that's the position of any geneticist. All I was saying was that when you look at something like eye color in *Drosophila* (or indeed in humans), although there are hundreds of genes and environmental factors that enter into making an eye, nevertheless if there's a one gene difference at a particular locus between two individuals, that can be the determining factor as to why that particular individual has a pink eye rather than a black eye. That is just straight genetics. You can't get away from that. The reason why left-wing ideologues attack books like *The Bell Curve* has nothing to do with that. Those critics are concerned with issues like race and I never mention race. I don't mind mentioning race, but it has nothing to do with what I was talking about.

Miele: But those same left-wing critics also attack your books.

Dawkins: But they don't read them!

Miele: Well, you document examples of such attacks in the end notes to the 1989 edition of *The Selfish Gene*. Isn't it just as ideological for Marxism to be brought in as an argument against genetic differences as it is to bring in Biblical fundamentalism as an argument against Darwinian evolution? Aren't we dealing with religion, rather than science here?

Dawkins: I suspect that we may be. The reason I think so is that the criticisms in some cases just seem to be silly. They seem to be a hysterical reaction to a misunderstanding. It's as though some people think that any mention of genes in the context of human behavior is somehow tainted with the tar of Social Darwinism and all the horrors that social scientists see in the history of their subject. So instead of just calmly and peacefully sitting down and thinking about what actually is the truth—"Are there genes that influence behavior?," the immediate response is to flame up old fires of what once upon a time were important political issues. That kind of thing bores me rigid! I care about what's actually true.

Miele: In a Darwinian sense, isn't it somewhat meaningless to argue about any supposed displacement of "superior" beings by "inferior" beings, or that evolution "is going backwards." Don't such arguments turn Darwinism on its head?

Dawkins: Because whatever evolves is, by definition, superior? There's nothing nonsensical about saying that what would evolve if Darwinian selection has its head is something that you don't want to happen. And I could easily imagine trying to go against Darwinism. I don't see why that's inconsistent. I can easily imagine saying that in a Darwinian world, the fittest, by definition, are the ones that survive and the attributes that you need to survive in Darwinian sense are the attributes that I don't want to see in the world. I can easily see myself fighting against the success of Darwinism prevailing in the world.

Miele: Shortly after publication of *The Selfish Gene*, you wrote a letter to the editor of *Nature* [the leading British science magazine, similar to *Science* here in the US], in which you stated that kin selection theory in no way provides a basis for understanding ethnocentrism. You said you made this statement, in part at least, to counter charges that were being made in the UK at that time by Marxist critics that Selfish Gene Theory was being used by the British National Front to support their Fascist ideology. In retrospect, do you think you went too far in trying to distance yourself from some would-be and very unwanted enthusiasts, or not far enough?

Dawkins: As to distancing myself from the National Front, that I did! The National Front was saying something like this, "kin selection provides the basis for favoring your own race as distinct from other races, as a kind of generalization of favoring your own close family as opposed to other individuals." Kin selection doesn't do that! Kin selection favors nepotism towards your own immediate close family. It does not favor a generalization of nepotism towards millions of other people who happen to be the same color as you. Even if it did, and this is a stronger point, I would oppose any suggestion from any group such as the National Front, that whatever occurs in natural selection is therefore morally good or desirable. We come back to this point over and over again. I'm definitely not one who thinks that "is" is the same as "ought."

Miele: How do you evaluate the work of Irenäus Eibl-Eibesfeldt, J. P. Rushton, and Pierre van den Berghe, all of whom have argued that kin selection theory does help explain nationalism and patriotism?

Dawkins: One could invoke a kind "misfiring" of kin selection if you wanted to in such cases. Misfirings are common enough in evolution. For example, when a cuckoo host feeds a baby cuckoo, that is a misfiring of behavior which is naturally selected to be towards the host's own young. There are plenty of opportunities for misfirings. I could imagine that racist feeling could be a misfiring, not of kin selection but of reproductive isolation mechanisms. At some point in our history there may have been two species of humans who were capable of mating together but who might have produced sterile hybrids (such as mules). If that were true, then there could have been selection in favor of a "horror" of mating with the other species. Now that could misfire in the same sort of way that the cuckoo host's parental impulse misfires. The rule of thumb for that hypothetical avoiding of miscegenation could be "Avoid mating with anybody of a different color (or appearance) from you."

I'm happy for people to make speculations along those lines as long as they don't again jump that "Is-Ought" divide and start saying, "therefore racism is a good thing." I don't think racism is a good thing. I think it's a very bad thing. That is my moral position. I don't see any justification in evolution either for or against racism. The study of evolution is not in the business of providing justifications for anything.

Miele: In *The Extended Phenotype* you talk about the Green Beard Effect, and last night Napoleon Chagnon, the President of the Human Behavior and Evolution Society, actually dyed his beard green before introducing you. What is the Green Beard Effect and why do you mention it only then to dismiss it as being too improbable to be a factor in evolution?

Dawkins: I use the Green Beard Effect as a way of explaining kin selection. If you imagine a gene that has two pleiotropic effects that apparently have nothing to do with each other (and in practice that's common enough), if one of its effects is to give somebody a label, such as the Green Beard, and the other is to give somebody a propensity to act altruistically towards individuals so labeled (that is, Green-Bearded individuals), then theoretically that gene will spread. The Green Beard Effect is a way in which a would-be selectively altruistic gene can recognize copies of itself in other individuals. That means that the gene can propagate itself by looking after copies of itself when it has the opportunity to do so. That's relatively easy to understand. But as far as I know, in practice Green Beards don't exist. But kinship is a kind of statistical Green Beard. Although your brother is not guaranteed to contain the gene that's making you practice "fraternal" behavior towards him, the odds that he has that gene are statistically higher than the odds that a random member of the population has it. Kinship is therefore a statistically watered down version of the Green Beard Effect that actually works.

Miele: Could there be selection for a mechanism that would operate like this —"Those who look like me, talk like me, act like me, are probably genetically close to me. Therefore, be nice, good, and altruistic to them. If not, avoid them?" And could that mechanism later be programmed to say "be good to someone who wears the same baseball cap, the same Rugby colors, or whatever?" That is, could evolution have a produced a hardware mechanism that is software programmable?

Dawkins: I think that's possible.

Miele: In his book, *Complexity*, Roger Lewin includes interviews with a number of evolutionary biologists. One of the topics they examine is progress. Is there a general tendency towards progress in terms of say, increasing neural complexity, increasing brain size, or increasing behavioral plasticity over the course of evolution? Or is "progress," as Stephen Jay

Gould has termed it, a rather noxious concept that we should read out of evolutionary thinking?

Dawkins: I think that there has been an almost hysterical over-reaction against the concept of progress. I've been as against some mistaken interpretations of the concept of progress as anybody. I very, very strongly object to the idea that living creatures can be arranged on a ladder, a kind of phylogenetic scale, with humans at the top. Not only should we not treat humans as being on the top, we should not see the animal kingdom as being layered as we often do. All zoology textbooks present their chapters in the same order—you start with protozoa, then you move through the coelenterates, then the flatworms, then the round worms and so on. Certainly that interpretation of progress is just a logical error. Evolution is a branching tree and that's all there is to it.

However, it is another matter to say that there is no progressive evolution within one lineage as you go from the distant past, through the more recent descendants, up to the present. There very well could be such progress. To the extent that adaptation is to the a-biotic environment (such as the weather), you would expect no progress. Evolutionary change would simply track the weather. If it gets cold, you get a thick coat. If it gets hot, you shed your thick coat, and so on. To the extent that adaptations are to the biotic environment (that is, other organisms, rather than natural conditions), then it seems to me quite plausible that there is in fact a progressive arms race as I term it. The better a predator gets at running down prey, the more it pays the prey to shift resources into anti-predator adaptations and out of other aspects of life. There are always trade-offs in the economy of life. If the predators are getting really good at their job, it makes sense for the prey to shift resources (which admittedly could have been put into making more offspring), into making better legs for running or better sense organs for detecting the predators. The predators then shift their resources accordingly.

Given the extraordinary elegance and beauty and complexity of the adaptations that we see all around us in living creatures, I think it's ludicrous to deny that those are the result of progressive evolution. There has been progress.

Miele: In your most recent book, *River Out of Eden*, you try to clear up some misunderstandings about the First Mother and the First Father. Would you like to repeat them?

Dawkins: I refer to things like the belief that Mitochondrial Eve was, like the mythical Biblical Eve, the only woman on Earth. Nonsense, she could have been the member of a huge population. She's simply the common ancestor of all living humans. Another error is to think that Mitochondrial

37

Eve is our most recent common ancestor. She most certainly is not our most recent common ancestor. That distinction much more likely goes to a male. The reason for that is pure logic and it's spelled out in *River Out of Eden*.

Miele: A few years ago, you and Stephen Jay Gould got in to a bit of an intellectual row about the question of punctuated equilibrium. By the time you wrote *The Blind Watchmaker* you seemed to say that more was made of the controversy by journalists than was warranted. What is your current position on this controversy?

Dawkins: I think that punctuated equilibrium is a minor wrinkle on Darwinism, of no great theoretical significance. It has been vastly oversold.

Miele: Why?

Dawkins: That's a matter of individual psychology and motivation and not my province.

Miele: You also took a bit of flak for likening religion (I think specifically Catholicism) to a virus? Is that still your position?

Dawkins: Yes. I come to it through the analogy to computer viruses. We have two kinds of viruses that have a lot in common—namely real biological viruses and computer viruses. In both cases they are parasitic self-replicating codes which exploit the existence of machinery that was set up to copy and obey that kind of code. So I then ask the question, "What if there were a third kind of milieu in which a different kind of self-replicating code could become an effective parasite?" Human brains with their powerful communication systems seem to be a likely candidate. Then I ask, "What would it feel like if you were the victim of a mind virus?" Well, you would feel within yourself this deep conviction that seems to come from nowhere. It doesn't result from any evidence, but you have a total conviction that you know what's true about the world and the cosmos and life. You just know it and you're even prepared to kill people who disagree with you. You go around proselytizing and persuading other people to accept your view. The more you write down the features that such a mind virus would have, the more it starts to look like religion. I do think that the Roman Catholic religion is a disease of the mind which has a particular epidemiology similar to that of a virus.

Miele: But couldn't the Pope (or Evangelical Protestants for that matter), reply, "Look, we just have a terrific meme. It's winning what you would describe as a Darwinian battle and you're angry because you just don't like it."

Dawkins: Religion is a terrific meme. That's right. But that doesn't make it true and I care about what's true. Smallpox virus is a terrific virus. It does its job magnificently well. That doesn't mean that it's a good thing. It doesn't mean that I don't want to see it stamped out.

Miele: So once again the discussion goes back to how do you determine whether something is good or not, other than by just your personal choice?

Dawkins: I don't even try. You keep wanting to base morality on Darwinism. I don't.

Miele: Given the number of popular and scientific controversies in which you have been involved, is there anything on which you have changed your mind, on which you'd like to correct the record, or which you see differently now than you did when you first became a household name?

Dawkins: Well, my second book, *The Extended Phenotype*, published in 1982 and summarized at the end of the second edition of *The Selfish Gene*, does downplay the role of the individual organism as the only vehicle of the genetic replicators. Previously, I rather blurred the distinction between vehicle and replicator. In *The Extended Phenotype*, I emphasize the distinction. I think I was right to say that fundamentally what is going on in natural selection is the differential survival of replicators. Replicators survive by virtue of their phenotypic effects upon the world. As it happens, it is a contingent fact that most of those phenotypic effects tend to be bound up in the vehicle (the particular individual body) in which the replicator (the gene) is housed. But that doesn't have to be so. I use the didactic device of looking at animal artifacts like beaver dams, the effects of parasites on hosts, animal communications, and indeed all of the interactions in ecosystems as illustrations of the ways in which genes might in principle and sometimes in fact do insure their survival by exerting phenotypic effects outside of the body in which they reside. That's a kind of change of mind.

My original purpose in introducing the concept of memes really was not to produce a theory of culture, but rather to say that Darwinism doesn't have to be tied to genes. It can work wherever you have a self-replicating code. We should actively be looking around for other examples of self-replicating codes which are "doing the Darwinian thing." The important thing is not to get too hung up on genes when you're doing your evolutionary biology.

Miele: What are the areas in biology into which Darwinism should be extended?

Dawkins: Sex is one that is being actively worked on. That is, the question of why it exists.

Then there's the embryological gap. In our Darwinism we postulate that there are genes for this and genes for that. We just leave the embryological causal link between genes and phenotype as a black box. We know that genes do, in fact, cause changes in phenotypes and that's all we really need in order for Darwinism to work. But it would be nice to fill in the details of exactly what goes on inside the black box.

To me, human consciousness is a deep, philosophically mysterious manifestation of brain activity and is in some sense a product of Darwinian evolution. But we don't yet really have any idea how it evolved and where it fits into a Darwinian view of biology. I don't know whether it will yield to a sudden flash of enlightenment, whether it will become one of those rather messy problems that never really get a proper solution, or whether it will eventually turn out that there never was a problem at all and that we were actually making up problems where there really weren't any. From where I sit, it seems to be a deeply difficult problem that has always been a philosophical problem but which I think is ripe for a take-over by evolutionary biology once we think how to do it.

Miele: Thank you very much.

GLOSSARY

ALLELE(S)—the alternative forms of a gene that can exist at a particular locus. Thus, A, B, and O are the alleles of the ABO blood group system; positive and negative are the alleles of the Rh system.

GREEN BEARD EFFECT—a term coined by Dawkins for the situation in which a gene has two effects (pleiotropy), one of which produces a recognizable phenotypic trait (the hypothetical Green Beard) and the other produces the tendency to manifest altruistic behavior toward others who also manifest that trait.

GROUP SELECTION (THEORY)—the theory that natural selection also operates at the level of the group and not just the individual organism.

INCLUSIVE FITNESS—the total reproductive success of an individual, including the proportionate success of his relatives based on the percentage of genes they share in common (100% for identical twins, 50% [on average] for siblings, parents, or offspring, 12.5% for first cousins, and so on).

KIN SELECTION (THEORY)—the selection of genes so as to cause individuals to favor their close genetic relatives as they are statistically likely to share genes in common. Often invoked to provide a neo-Darwinian explanation of behaviors such as altruism.

MEME(S)—a term coined by Dawkins for units of cultural inheritance, analogous to genes (the units of genetic transmission), which are acted upon by natural selection.

MITOCHONDRIA—the cell organelles that are the site of energy-releasing biochemical reactions. Mitochondria have their own DNA, which is passed through the female line only and thus is often used in tracing genetic lineages. **MITOCHONDRIAL EVE**, believed by some to be the common ancestor of all humans, was determined by analysis of mitochondrial DNA.

PHENOTYPE—the observed trait, morphological or behavioral, manifested by an organism. It is the product of the organism's genotype and the environment in which the organism has developed.

PLEIOTROPY (-IC, -ISM)—the condition in which one gene produces two (or more) different phenotypic traits.

PUNCTUATED EQUILIBRIUM (THEORY OF)—the theory proposed by Stephen Jay Gould and Niles Eldridge that evolution can consist of long periods of stasis (no change), punctuated by short periods of rapid change (speciation events), rather than a steady series of small, incremental steps.

CHAPTER 3

THE IMPERIAL ANIMALS
25 YEARS LATER

AN INTERVIEW WITH LIONEL TIGER AND ROBIN FOX

Lionel Tiger and Robin Fox first met, appropriately enough, while attending a symposium at the London Zoo. In 1971 they published the first edition of *The Imperial Animal*. The wealth of new information that had recently become available from zoology, biology, history, and genetics, Tiger and Fox argued, demanded "a skeptical review of how we explain our political behavior, our economies, our leisure, our forms of education, and the recurring tragedy of war." An augmented edition of *The Imperial Animal*, with a 1973 introduction by Nobel Prize-winning ethologist, Konrad Lorenz, was published in 1989.

I trapped Tiger and Fox recently in Princeton, New Jersey, and asked them to look back on the 25 years since the publication of the first edition of *The Imperial Animal*, and assess the impact of Darwinian thinking on the social sciences.

Robin Fox is also the author of *The Challenge of Anthropology* (Transaction, 1994), *Encounter with Anthropology* (Transaction, 1980), *Kinship and Marriage* (Cambridge, 1984), *The Red Lamp of Incest* (University of Notre Dame Press, 1983), and *The Search for Society: Quest for Biosocial Science and Morality* (Rutgers, 1989). Lionel Tiger has also written: *Women in the Kibbutz* [with Joseph Shepher] (Harcourt, Brace, Jovanovich, 1975), *The Manufacture of Evil: Ethics, Evolution and the Industrial System* (Harper and Row, 1987; M. Boyars, 1991), *Men In Groups* (Random House, 1969; M. Boyars, 1989), *The Pursuit of Pleasure* (Little, Brown, 1992) and *Optimism: The Biology of Hope* (Simon & Schuster, 1978; Kodansha, 1995). Tiger's books have been translated into 10 languages. Tiger and Fox shared the position of Research Director of the H. F. Guggenheim Foundation from 1972 to 1984, and both presently teach anthropology at Rutgers.

Figure 3.0: Robin Fox & Lionel Tiger

Miele: It's been 25 years since you published *The Imperial Animal*, one of the first books in the period following W.W.II to examine human social behavior from an evolutionary perspective. Since then, there's been an explosion in evolutionary biology and evolutionary psychology. Looking back, where were you right and where were you wrong, and where do you two disagree?

Tiger: We were right in obeying the Law of Parsimony. What we sensed intuitively and demonstrated through the materials we used, was that the conventional wisdom as epitomized by the gaseous penumbra that was the social sciences at that time couldn't effectively explain some very basic issues. The strength of the Darwinian approach is that it is both wondrous and empirical at the same time. You can look at a flower and marvel at how beautiful it is and you can also understand how it works. We were observing this complex, wonderful species—*Homo sapiens*—with wonder in our eyes, while at the same time being very practical and empirical about it.

Fox: We were theoretically right and have been proved to be so, but we were tactically wrong. We had this, I now think foolish, notion of trying to convert the social scientists to our point of view. If anything, the current divide in the social sciences between the anti-Darwinian, hermeneutic, postmodernist type of explanation, and the essentially scientific model of explanation that we favored is worse than it was when we wrote the book—the other side has taken over the social sciences.

What we should have done was just set up a separate organization with a separate journal and bypass the traditional social sciences altogether. We had an almost pathetic sort of belief that sociology and anthropology could be reformed. We were naive about many things, including human nature!

Tiger: I don't know where Robin and I disagree. Now when I look at the book, I just don't know who wrote what. I'm sure Robin doesn't either, though he will undoubtedly claim that all of it is his! The fact is that we wrote it as a complete collaboration. I don't discern, apart from some crankiness he has about some political matters, any serious intellectual differences between us as to the actual science involved. And that science is really thrilling.

Miele: A major criticism of all such evolutionary attempts is that they are reductionist. You are both social scientists. What is the role of reductionism and of emergence in properly explaining human social behavior?

Fox: Reductionism has got an unfortunate bad name.

Tiger: In fact, reductionism is a good thing.

Fox: Successful sciences have done precisely that. In the natural sciences, the more you can reduce things to their physical properties, the more successful you think you are being. Insofar as you can reduce explanations to simpler properties, that's fine.

I think the problem is that there have always been two sets of people coming at this problem: one bunch from the social sciences (with us as the cheerleaders, as it were, from the beginning), and another crowd from the natural sciences (biology, zoology and so on), with primatology sort of sitting in the middle somewhere and the anthropologists sometimes having a foot in each camp. The ones coming from the natural sciences tend to be more reductionist in the sense that they are more interested in ultimate explanations.

People often ask, "What's the difference between you two and someone like E. O. Wilson" (who coined the term "sociobiology" and wrote that magnificent compendium on it). I think the difference is that sociobiologists are much more interested in some ultimate evolutionary explanation and why certain ultimate evolutionary mechanisms exist. We, on the other hand, were coming from the social sciences and addressing most of our arguments to our colleagues in the social sciences. There you have to deal with people who are satisfied with proximate explanations and individual motivation.

The natural scientists then say, "This is really just an instance of someone maximizing his reproductive success." That, in turn, elicits a tendency on the part of the social scientists to resist such reductionism. We've been caught in the middle. The great difficulty is getting the two sides to see that ultimately the two methods are not incompatible, though they can look wildly different when approached from such different angles.

Tiger: It's also often difficult for social scientists to see behavior as natural. In the universities you have natural sciences and social sciences, as if these were two separate realms, which in turn presupposes that social behavior is somehow non-natural. But, social behavior is natural.

I remember very early on, Robin telling me that one of the two things he learned in graduate school was the principle of the normal curve.

Fox: I learned about the normal curve in undergraduate school. I didn't learn anything in graduate school!

Tiger: Sorry. If you understand that all life operates on a normal curve, then you know that there is going to be variation and there is going to be centrality within that variation. In an unusually good phrase of his, Robin talked about "ethnographic dazzle"—the temptation on the part of some social scientists to look around the world and see all these variations in some behavior and conclude that there is nothing in the center. Well, if you statistically graph that behavior, you find that there is a central tendency, just as there is with physical characteristics such as height, skin color, blood type, etc.

Somehow, social scientists have a problem dealing with their own literal reality, namely that social behavior is natural. Instead, they often try, in ways reminiscent of theologians of past days, to see something else than what's really there. Rather than invoking God as the supernatural entity, they invoke something they call "Culture." The two concepts are equally confused and oceanic. But it's all just one more attempt to make that old Lévi-Straussian distinction between nature and culture.

Miele: Another major criticism is that evolutionary biology and evolutionary psychology consist of "just so" stories. To what extent can we actually test hypotheses in evolutionary biology? Is it science on a par with physics or is it more like cosmology where we first guess what must have happened and then extrapolate what ought to have happened?

Tiger: Joseph Shepher, Robin's student, and I did a study of the Israeli kibbutz movement which we published in a 1975 book called *Women in the Kibbutz*. We looked at what happens when you fundamentally change all the conditions of society—men and women both work, they have the same income (which is, in fact, no income), the kids are raised in a children's house, and so on. Based on the mammalian ethogram, we predicted you would find that eventually this system would not work and, if nothing else, mommies and babies were going to want more proximity and contact than they were allowed by the male ideologists who set up the whole system. And that is, in fact, what we found. So that's not a "just so" story; it's a "just is" story—it's what really happened. What we've been able to do is to reduce the degrees of freedom needed to approximate human nature.

Fox: A lot of the criticism of sociobiology as "just so stories" is justified. That is, if you find a certain correlation between things in the world, then it's relatively easy to come up with an evolutionary principle to explain it. For example, "Why do women have orgasms?" There have been a parade of answers. In fact, in certain areas of the world where speed of intercourse is considered macho, and therefore very few women ever have orgasms (unless they're self-induced), the fertility rate is highest. So there is no correlation between the state of female orgasm and the reproductive rate. My own hypothesis is that orgasm is a by-product—a total accident—and that it's not adaptive at all.

Stephen Jay Gould, among others, has led a charge against sociobiology on the grounds that it is pan-adaptive—whatever exists, exists for an adaptive reason. Clearly this is not so. A great deal of behavior is either maladaptive or neutral. There is a tendency, once you get on the evolutionary bandwagon, to find an evolutionary explanation for everything. I don't think that is bad as a first approximation. Your first step is to wonder what the evolutionary explanation might be. But then you've got to move beyond that to properly testing hypotheses and not simply making up "just so stories."

Tiger: But wait a minute. Even on the business of female orgasm, we now have some data that do indeed show its function. The film by Robin Baker and Mark Bellis, used so effectively in the Desmond Morris video series, *The Human Animal*, shows that at the point of orgasm, the woman's muscular contractions cause the tip of the cervix to dip down into the pool of spermatozoa. So you have an actual abetting of fertility in that the spermatozoa have "an easier ride." There is an evolutionary mechanism here and I suspect we might find physiological mechanisms that underlie a whole series of other behaviors that we might have thought were just whimsical or random by-products.

Miele: But one might ask what have evolutionary biology and evolutionary psychology taught us that our grandparents didn't already know, even if somewhere along the way, we and our parents forgot it? A lot of evolutionary biology seems to be simply confirming the obvious.

Tiger: If my grandfather had been a political theorist and if he had written a book called *The Inevitability of Complete and Identical Equality*, I might have accepted his argument at one time, but now I wouldn't. Now we know that humans tend to create hierarchies and you have to work damned hard to prevent them from emerging.

Fox: People of my grandmother's age and generation were full of all kinds of folk wisdom—some of it was cock-eyed and some of it was spot on. What they all accepted was that there is a human nature. "There's no changin' human nature, lad!" was one of my grandmother's constant refrains that I still remember. But this human nature was largely a collection of their folk beliefs. What they did know was that there was a human nature and that indeed it is hard to change. But they didn't necessarily know the role of evolution in producing it. What they in their folk wisdom knew is at the root of what we are saying about human nature. But then you have to discover exactly what that is.

Tiger: Our grandparents' generation did have a kind of grainy sense of the nature of family and kinship, which in the generation you talked about, got lost in a swamp of therapy. In a curious way, what the evolutionary biologists have really done is to try to get back out of that swamp to the set of irreducible behaviors that generate the emotions and passions in the first place.

Miele: A major battle flared up recently following the publication of the controversial book *The Bell Curve*, especially as regards its treatment of the issue of race. We organized a Skeptic Symposium at Caltech and an issue of the magazine (see Chapters 10, 11, and 12) to it. Do either or both of you have any comment on the current status of the heredity-environment question, the scientific status of the concept of race, or the meaningfulness of IQ?

Fox: Here we disagree!

Tiger: And we're prepared to talk about it.

Miele: Good. That's just what *Skeptic* wants.

Tiger: Do you want to go first, Robin?

Fox: You go first; I'll be the skeptic.

Tiger: I once advocated at a meeting of the American Association for the Advancement of Science that IQ tests be put under the statutory rape legislation so that nobody under the age of 18 could be required to take one. We're doing this interview here in Princeton, NJ, right near the luxurious headquarters of the Educational Testing Service, which, if you drop by, even though it's zoned as a farm for tax purposes, represents the dream of the American upper middle classes—trying to figure out ways to test people to insure that the right people get into universities.

If you look at the history of IQ tests, and there are a couple of recent pieces by Lemann in *The Atlantic Monthly* about this, you see how class-based they are. There was also some eugenic notion, driven in part by the war effort, that quality will out and that there was a some special virtue associated with doing well on these tests.

First, I remain very doubtful of tests because of how they are created. They invariably remove questions that distinguish by sex, for example, which means that what you're getting is a kind of unisex haircut. This may in fact remove the questions that distinguish between males and females that are sharply rooted in their maleness or femaleness. This, in turn, could be one of the main reasons that you get differences between blacks and whites. In effect, you pull out the questions that reflect various aspects of bipolar social, sexual competence. Further, it doesn't seem to me that the assumptions made in producing tests justify the gravity of the conclusions in how they are used. As to heredity and environment, both are completely essential to any organism.

One thing that I think is a dismal consequence of this IQ business (the SATs and so on) is the emphasis on math. It seems to me that if any single human skill has the least possible thing to do with evolved capacities of a Paleolithic hunter-gatherer, it's advanced math. What we have done is to turn over the elite positions in our society to those people who can best do advanced math. The French are now beginning to realize this. They have a completely exam-oriented educational system. As an engineer who runs a large enterprise in France told me, "I can't get my people to non-optimize." So the French continue to make these ridiculous products which nobody will buy because they are perfect, except that they don't work in human terms.

Fox: Lionel has this crusade against IQ tests.

Tiger: I only do it to amuse Fox!

Fox: IQ tests have never bothered me. I've always found them rather fun. There's no question that IQ tests are highly predictive of success in our kind of intelligence-oriented world. There's also no question that a large amount of IQ is inherited. Twin studies are now into the thousands. Probably all the identical twins they can find in the world have now been tested. These studies show beyond doubt that a large percentage of IQ is inherited. But what I think is not taken into account is that even if a large percentage of IQ is inherited, even as much as 80%—and the estimated heritability varies between 50% and 80%—that leaves between 20% and 50% that isn't inherited. What's very rarely been studied is what the environment contributes to IQ. We understand the genetics and the inheritance of IQ very well. What we don't understand is the nature of the environmental contribution.

Native American children do well on IQ tests, even though they come from what would be described by many sociologists as very deprived environments. They often don't have running water, electricity, 24-hour television and so on. But having lived among these people as I have, one realizes the tremendous stimulation and input that goes into the upbringing of a young Native American child—the enormous amount of training in ritual and singing in the Pueblo children I saw—they grow up in an incredibly rich environment: an environment that a lot of white kids in Princeton sitting in day care centers waiting to be picked up by their parents would envy. This is not a function of running water or electricity. It has to do with the amount of care and attention and intelligence that the parents themselves put into bringing up their children.

There are significant differences in the average IQ of different races. But the distributions overlap so enormously that we've no idea really what part of mean difference and of the overall variance is the result environmental input. If it's 50-50, the environmental component could be responsible for all of the mean difference and genetics may have nothing at all to do with it. Even though we know that a large part of IQ is inherited, it doesn't necessarily follow that the mean difference between races is the result of genetics. What isn't properly understood is the contribution of the environment to the growth of the brain and the development of intelligence.

Tiger: I was just at a meeting in Vienna in June of scholars from around the world, including Cavalli-Sforza from Stanford, who argued that you simply can't use race as a concept. We now know that we all probably evolved from one group of people who lived about 200,000 years ago. Unfortunately, the simple-minded notion of heredity versus environment has made this ugliness in the discussion of race and IQ possible. Robin once wrote a piece for *The New York Times Magazine,* "The Chinese Have Bigger Brains. Are They Superior?"

Fox: That was their title, not mine! My title was, "The Abolition of Race." It was really an article advocating widespread miscegenation. I pointed out that race was a very useful episode in human evolutionary history. I agree with Lionel that race is not a very useful scientific concept. We know roughly what we are talking about—skin color, hair form, and so on, and there are rough divisions of mankind on this basis. But they are very rough and scientifically not very useful.

Racial differences were useful in an evolutionary sense in the period when the human species was spreading out all over the globe. These minor adaptations, which is what they are, helped us to adapt and spread into numerous micro-environments. Now when we have clothing, buildings, artificial climate control, these physical adaptations just don't mean very much. Individuals live all over the globe regardless of their skin color or hair form. We just don't really need these adaptations anymore.

But I did suggest that if we were to expand to other planets and environments, then race and racial differences could become important again. For example, if people were to live on Mars, it might be helpful to their survival if they had a particular type of bodily form that would be different from those who lived on earth or who might try to live on the moon. Genetic variation could again then be useful. At the present moment on Earth, however, there's scarcely any use for it at all. So I suggested that we simply breed out this variation so that everybody would probably end up with a good tan and resistance to epidemic diseases as a consequence.

Needless to say, I got hate mail from both sides. The Black Power people were going to have my guts for garters because they thought I was suggesting that we breed out blackness. And you can just guess what I got from the White Power people.

Miele: If race is not a very meaningful concept, why do we see America so deeply racially polarized 30-odd years after the heyday of the Civil Rights Movement, as President Clinton had to acknowledge recently?

Fox: When we say race is not a useful concept, we're speaking in scientific terms. It's not a very useful concept.

Miele: Then why is it so salient at the gut level?

Fox: Because of a persistent factor of human nature that our grandparents knew all about — xenophobia! We have a deeply built-in fear of the stranger. This is part of a Paleolithic spacing mechanism. Tribes were separated in space and there were some individuals that were like you and some that were not like you, who were by and large not well disposed to you. Therefore, we have a similarity detection mechanism built into us.

51

From childhood, we tend to develop a picture of an ideal form or face from the observation of the people around us. We have a special part of the brain that sorts through faces looking for familiarity. Those that are least familiar are those that are going to be the most frightening.

And even if nature doesn't provide the cues to familiarity, like skin color, for us, we provide it for ourselves with things like costumes, haircuts, tattoos, headdresses, things through the nose, or anything that distinguishes who we are from who they are. Skin color is merely one aid to this inborn xenophobia. Something deep down in that Paleolithic brain, registers "Different, Different, Different!"

Tiger: I agree, but I defy either of you to tell me the observable racial difference between a Serb, a Croat, and a Bosnian Muslim, for example.

Fox: As I said, if there isn't an observable difference to separate the in-group from the out-group, we'll make one up.

Tiger: So we make up religious categories, such as Muslim versus Catholic versus Orthodox, or whatever. People appear to find such differentiation easy to learn and are then quite willing to find their group superior to the other group, and to then find it necessary to "defend" their group against the other.

Miele: Somewhere in your book I encountered the words, "The political process—the process of redistributing control over resources among the individuals of a group—is a breeding process. The political system is a breeding system" (p. 25). Does either of you want to take credit (or blame) for having penned those lines?

Fox: I don't know who actually penned them. Either of us could have. We were very interested in looking at primate breeding systems. I think I was the one who invented the term, "the primate baseline." The idea was that since we evolved from some kind of primate, if we look at enough primates to see what we all have in common, we'll get some idea of what it was we evolved from. If you see what we had to change from to get to be what we are now, it might help to explain what we in fact are. You do run into the danger of invoking "just so stories." But if you do it right, you can avoid that pitfall.

The important thing about all primate breeding systems is that they are arranged hierarchically—the top males get to do the majority of the breeding when the females are ovulating. We were interested in finding the extent to which this was still true for human breeding systems. Humans, of course, introduce politics into the process. Non-human primates have a crude sort of politics in which some males gang up to displace other males. Humans, obviously, have true political systems and some males get to the top of them.

If we were behaving in true primate fashion, the point of getting to the top of the pyramid and being in charge would be precisely that you get to breed with the most females. In the vast majority of political systems over the course of world history until relatively recently, this has indeed been the case. One can rattle off all the examples. Laura Betzig wrote an excellent book, *Despotism and Differential Reproduction*, on the subject, in which she went through all the ancient civilizations showing the huge harems and the massive numbers of offspring produced by Solomon, Pharaoh Ramses II, or Ismail the Bloodthirsty of Morocco.

But what happens in those systems where you get to the top of the pyramid and this reward doesn't materialize? Jimmy Carter lusted in his heart, but I don't think there's any evidence that he lusted in the backstairs closet like Warren Harding. I'm curious as to where the energy and the imagination and so on of the males who get to the top of the hierarchy goes in such cases. Their entire physiology is probably telling that they should breed like crazy with every female they can possibly control. A lot of our troubles are probably the result of the fact that we don't provide our presidents with harems. Maybe if they had to spend a lot of time having to deal with the quarrels in the harems, they'd keep out of our lives.

Miele: Some would argue that the JFK administration tried to do exactly that.

Fox: It got close. Quite close!

Miele: I want to turn to the most recent intellectual counteroffensive against evolutionary biology, namely the argument from deconstructionism and literary criticism that everything, including science, and especially the stuff you fellows have been telling me for the last 45 minutes about male and female behavior, is all just text that must be interpreted. You've both written and spoken on this subject, so have at it.

Tiger: As Alex Argyros has argued, this kind of hermeneutic hash is a form of Creationism without God. That is, everything after the Deluge or the Great Event is merely arbitrary and that there is no empirical, central basis for anything.

Fox: The argument itself seems so self-evidently stupid that the more interesting question to me is the sociological one: why has such a view spread like wildfire through North American universities? It's already dead in Europe.

Miele: Even in France, where this all started?

Fox: Yes, it started there, but the French rapidly get bored and move on to something else.

Tiger: After all, they're responsible for giving us the world of fashion.

Fox: In Frankfurt, being German, they go solidly on for a lot longer. But in Paris, that was last year and they're interested in this year's trend.

Miele: One explanation for the spread of postmodernist philosophy is that it provides the literary mindset a way to strike back at science.

Fox: True. The content of it is so clearly rubbish it's as if you're suddenly faced with entire geography departments going over to Flat-Earthism. What is extraordinary is how they were able to do this and why.

Whenever there's a question about ongoing social policies and issues in this country that I don't quite understand, I ask myself how the baby boom is related to it. More often than not, it lurks behind and explains a lot of what is otherwise unintelligible. In this case, you have to realize that the structure of American universities is such that people come into them from high school semi-literate. This doesn't usually affect European universities, where the students are highly selected and they're quite literate when they come in. But in American universities, the first year is basically a remedial operation in which you desperately try to teach the students to read and write. So English reading and writing courses have become compulsory for everybody, and huge English departments have grown up. But they haven't been staffed with what was really needed, namely high school- and junior college-level remedial English teachers. They've been filled with young hopefuls from English departments who want to write their books on "The Use of the Semicolon in the Portuguese Sonnet," or something like that, and get promoted and become full professors.

So you get these enormous departments of English, full of all these aspirants, with not that much for them to do when it comes to English Literature, which is largely the spinning of opinions with the apparatus of scholarship around it. You end up with a body of very frustrated people, low on the academic pole, who have always felt themselves to be poorly regarded, as opposed to the natural scientists who get all the big, fat grants and all the prizes and recognition. Suddenly, this absurdly relativistic doctrine comes in from Paris that says that everything is just text. There isn't even an author anymore. The author was abolished. So only the texts and the critics remain. The critics, therefore, become the ultimate arbiters of everything and since it's all only opinion anyway, any opinion is a good as any other. This then gets elevated, with the help of a little nihilist philosophy, to a sort of metaphysical doctrine which they can all join in on. From being the down-trodden of the Earth, this horde of not very good or very smart remedial English teachers become the center of the intellectual universe. They then get to tell the scientists what they can and can't do, and what constitutes truth. And since, in this view, nothing constitutes truth, anything they say is as good as anything the scientists say.

The really unfortunate thing is that they're now tied in with the PC-crowd, the academic witch-hunters and generally leftist anti-scientific bunch who are the leftover Marxists (who also don't know what to do now). And so this strange gaggle of odd bedfellows are now running the show. They're swallowing up anthropology. You can't talk to anthropology graduate students any longer except in this peculiar sub-language.

Tiger: In particular, the notion that sex roles are completely arbitrary and that patriarchal, sexist patterns of social life have not only maintained the power system but have been responsible for all the intellectual, and more important, the scientific structures, is very strong. Therefore, you have many arguments, some skillfully done, showing that, for example, biology with its interest in reproductive investment and so forth, is a metaphor for capitalism and a male-centered process. These arguments are plausible on their face. They make a good sermon. But the underlying notion that supports them is that everything is the same and that all existing scientific structures, like all existing power structures, simply reflect the ideas of a bunch of largely white, power hungry men.

Fox: But they commit a serious logical error by totally confusing the practice of science with the truths of science as established by scientific procedures. Science may very well be dominated by males. In the Middle Ages it was dominated by Augustinian monks. Tomorrow, it could just as well be dominated by Tierra del Fuegian-speaking hunchbacks. The truth of Boyle's Law, or of $E = mc2$, is not a function of its provenance or its use. It may be used to make atom bombs and kill people or it may be used for totally benign purposes. But neither of those affects its scientific validity.

Tiger: Roy D'Andrade, in *Current Anthropology*, mockingly asks, "Isn't it odd that the person responsible for all the misery and inequality in the world happens to work in the office just down the hall—a sociobiologist or a geneticist, rather than empirical operators of the system itself?" This is really an internecine contest over status within the university hierarchy as well as an attempt to unseat the notion that you have to exert some effort in order to actually get it right. One of the reasons social scientists have been so eager to evict reductionism is that it requires them to learn something about the brain, genetics, or the behavior of other species. That's hard to do. If you can avoid having to learn all that simply by saying "it's reductionist," you're well ahead of the game.

Miele: Can we turn all this around and look at story-telling from Homer and the Bible to *Melrose Place* and *Murder One* in evolutionary terms?

Fox: Joseph Carroll has taken the lead in arguing that evolution gives you a reference point from which to judge literature. This, of course, is anathema to the relativists and the hermeneuticists. If you know something about the evolution of sexual selection and the role of female choice and the evolution of breeding hierarchies and how they've become instantiated in our own species, you can then look at literature in those terms. Epic literature and "the great novels" can be evaluated in evolutionary terms. The prime mover is usually some kind of sexual conflict for the fertile female between older, established males and younger males. In the *Iliad*, it's Agamemnon versus Achilles over Briseis; King Mark versus Tristan for Isolde,...

Miele: Am I venturing too far to say that's why our society has been so much more fascinated by O. J. Simpson, Nicole, and Ron Goldman rather than by the Savings and Loan crisis, which is costing us all a hell of a lot more money?

Fox: Yes — the Savings and Loan crisis doesn't ignite our limbic system. The Paleolithic didn't have Savings and Loan institutions, or anything even corresponding to them. But it did have sexual competition. So anything having to do with sexual competition—death, love, sodomy, incest, violence, you name it—hits our limbic system and sets off a response.

Tiger: While there may not have been Savings and Loans in the Paleolithic, there were malefactors and liars. I think humans in fact have a metaphorical gene (or genes) for morality. Living as we did in small communities, you would quickly identify the Charles Keatings. He would never have gotten the opportunity to swindle two billion dollars worth of meat, as the group would have quickly hacked his head off well before he got away with the first hindquarter! While the Savings and Loan scandal is not so inherently dramatic as *The People versus Orenthal James Simpson*, the commitment to root the scoundrels from our midst remains quite strong.

Fox: Yes, but we also make the scoundrels the heroes if they are essentially trickster figures. Trickster was one of Jung's great archetypes who shows up in folk tales and literature all around the world and down through the ages. We love characters like con men, provided, of course, the people they're taking richly deserve it and are scoundrels themselves. But we hate the traitor, the turncoat, the individual who betrays his own group. Those are the ones we really want to draw and quarter.

Miele: Speaking of evolution and morality, David Hume dismissed going from "Is" to "Ought" a long time ago. But does Darwinism present us a way of rising above the "Is-Ought" problem by telling us how it is that our sense of ought ever evolved?

Tiger: We can't necessarily go from "Is" to "Ought," but we can certainly go from "Is" to "Is." The Darwinian perspective is not necessarily, or even desirably, a set of moral recommendations. It is merely an empirical starting point that permits you perhaps to get the facts right. On that basis, you might then try to create some sensible social policies. For example, when Karl Marx and his friends created their vision of a utopian world based on a severe misunderstanding of *Homo sapiens*, they doomed half the planet to disaster for 70-some years. If Marx had actually paid attention to Darwin in some systematic and thoughtful way, rather than merely thinking of himself as the economic analog of Darwin, he might have concluded that you might want to have the family and the kinship structure still form the basis for a large part of the social system. The kinship structure is part of living biology.

Instead, Marx denied the power of the family altogether. The poor people in the kibbutz that I studied, literally believed that you had to have the children over six weeks old living in a children's house, some form of Newt Gingrichian barracks, in order to destroy the effects of the terrible, old social structure. If you don't understand the family nature of human beings, and opt instead for some kind of bureaucratic ideal of impartial non-emotionality, you're probably going to get it all wrong.

Miele: But one could argue that Marx did in fact identify sociality as a basic characteristic of human beings, whereas some sort of hyper-Adam Smith, free market approach would argue that we're all so atomized that we just don't care about things like family. One deal is as good as any other if the bottom lines come out the same.

Tiger: Adam Smith's most important book is *The Theory of Moral Sentiments*, though it seldom if ever gets identified. His famous economic treatise, *The Wealth of Nations*, is obviously important. But given the

Scottish Enlightenment context out of which he came, *The Theory of Moral Sentiments* is as important, if not more. In it, Smith says that you have got to satisfy all these social needs as well as the economic ones.

Fox: While we're talking of the Scottish Enlightenment...

Tiger: I knew this would turn Robin on.

Fox: ...so recently popularized by James Q. Wilson in his book, *The Moral Sense*, Hume's discussion of "Is" and "Ought" is often misconstrued. He was actually making a kind of anthropological observation. What Hume said is, "Isn't it interesting, when you listen to people making moral arguments, how frequently and easily they jump from 'Is' to 'Ought,' without giving you the intervening reasons." Hume himself was quite willing to jump from "Is" to "Ought." If you read Hume on morals, he does it all the time. He never says that you shouldn't or couldn't do it. Indeed, Darwin, in his chapter on moral sentiments in *The Descent of Man*, agrees with Hume. Hume has been misinterpreted by almost every philosopher since. I say this because this so-called Naturalistic Fallacy is often thrown up at us.

Further, much of social science has picked up on the Western utopian tradition so we forget that, until the Renaissance, there was no utopianism. Rather, there was the notion of a Golden Age that we had fallen from. There was no idea of Golden Age that we were going to achieve on Earth; maybe in the next world, but not here and now. Can you imagine a research proposal going to any social science research council or grant agency that said, "this research will statistically demonstrate that there is no possibility of ever improving the lot of mankind. Instead, I shall demonstrate how, over time, things will only get worse and worse."

Miele: Malthus tried that, didn't he?

Fox: Do you think he'd get a grant today? No, there's an automatic, built-in assumption that whatever we learn can then be used to make things better. But it's not an unreasonable assumption that you can't make things any better and that we are living in what is the best of all possible worlds, even with all its problems.

Miele: Something that evolutionary arguments seem to share with a good bit of left-wing and even deconstructionist philosophy is a tendency to attack, for lack of better words, traditional morality and religious institutions and practices. But doesn't it make sense from an evolutionary perspective to argue that, even if the theology as such is goofy, the religious practices and behaviors must have some positive survival value if they are still around after all these years? Hasn't a lot of evolutionary biology acted as a sort of intellectual "wrecking ball" that has knocked over some of the guard rails erected by traditional morality, thus throwing us on nothing but our own meager resources?

Fox: There have been interesting turns and twists in the history of the dialogue between Darwinism and religion. When Darwinism first appeared it was treated by the religious establishment of the day as the ultimate attack on religion, and its rise was thought to mean the destruction of the entire basis of the religious foundations of the moral order. This is still thought to be the case by creationists and right-wing religious people, as was seen in the 1920s in the Scopes trial. But there have been interesting twists in this tale. When Robert Ardrey wrote his interesting, if in some ways flawed, books, his arguments were often taken up by the religious crowd because they thought his work confirmed their views on original sin and the fallen nature of man. Some of the most enthusiastic supporters of his views on territoriality and aggression were bishops and other clerics.

You can take Darwinism in a number of ways. Darwin's Bulldog, Thomas Henry Huxley, followed it into agnosticism, as did Herbert Spencer. On the other hand, you can say that evolutionary theory suggests there is a need for myth and for some system of illusion by which to live. Carveth Read, in *The Origin of Man and His Superstitions*, in the 1920s talked about the need for illusion in the lives of early hunters. He saw that a great many things that rationalists were dismissing as mere irrational superstitions must, in fact, have performed some extremely important function for those people since they spent an enormous amount of time and effort on them.

Tiger: Even with contemporary and skilled biological analysis there's a real danger of looking at religion in a mocking sense and concentrating on its negative features, as exemplified in Richard Dawkins' opposition to any sacred element in the educational system and his view of Catholicism as a mind virus, or John Hartung's analysis of the xenophobic bloodthirstiness of the Bible. (see Chapter 2 and Hartung, 1995). I may agree with them personally, but speaking as someone who tries to be a natural scientist it seems to me that if one says that something that 99% of a population is doing is unnatural, you have a big problem. What I tried to say in my book *Optimism* is that whatever you think about religion, you've got to deal with it and explain why people are making these statements, why they are building these buildings, why they are giving 10% of their money to these characters. Just what the hell is going on here? I don't understand it half the time. A lot of it, as you so charmingly put it, is "goofy" and even beyond goofy; nonetheless, there it is.

Miele: Well, contrary to much of religion, evolutionary biology can end up making the dismal science of economics seem like a joy ride. If you aren't a beautiful, young female, or a powerful, high-ranking male, what has evolutionary biology got to tell you except that you're a loser? Doesn't evolutionary biology act as an acid eating away at the very basis of society?

Tiger: While your caricature of evolutionary biology is telling and amusing, I think at the same time we all understand that most of us don't fit the leading man or leading lady model. But we do what we can in our context with our limited beauty and power. And at the end of the day, we probably think we did OK. We got a smile from somebody we probably didn't deserve a smile from.

Miele: Does this then go back to your argument that the system is biased in an optimistic direction, and in fact it has to be in order for us to get through the day?

Tiger: I keep coming back to Isaac Bashevis Singer who, after he received the Nobel Prize in Literature, was asked if he believed in free will. He replied, "Of course I do. I have no choice!" And in a way, we have no choice but to assume that things are going to work out slightly better.

Fox: I think this bleak picture of the dismal science of evolutionary biology that you paint isn't the whole picture. There are other things that get investigated. In primate groups, the alpha male may sit there in the center, looking very lordly and seeming to get the majority of the females. But he's not necessarily the most interesting, nor the most active of these characters. And in some groups, he's not even the one who gets the majority of the females. There's a marvelous term of Clutton-Brock's to describe the junior or peripheral males in these hierarchical breeding systems, of which *Homo sapiens* is one: the Sneaky Fucker. While the alpha male is sitting in the center, looking out into the distance with lordly disdain, one of the junior males is going off with the females, who are often quite willing. The alpha male is often a very harassed, cuckolded character, who needs to spend a lot of his time looking over his shoulder.

Miele: Sounds like King Arthur, Sir Lancelot, and Queen Guinevere.

Fox: Exactly. Arthur is too busy presiding over his Round Table to see what's going on under his very nose. A lot of the characters that fascinate us, both in fiction and in real life, are Sneaky Fuckers. They're cunning and clever in playing what was perhaps a poorly dealt hand to such a smart advantage that they make the alpha male look dull and boring in comparison. There's a brighter side to your bleak picture for those who are ostensibly not at the top of the ladder. Perhaps we should write a book called *Alpha Males Don't Always Have All the Fun.*

Miele: Didn't *Animal House* prove just that?

Tiger: The great thing about the biological perspective is that it encompasses individuals on each side of the spectrum.

Miele: At *Skeptic*, we like to turn our skeptical eye toward extraordinary claims, often to religion or New Age thinking. But perhaps we're shooting at rather small fish in a tiny barrel. Maybe the proper object of skepticism should be what is taught in colleges and universities. How much do you think we should focus on the social science curricula currently taught in a major universities? Should we indeed be skeptical of the whole enterprise of social science?

Fox: Absolutely! We've spent our whole lives being skeptical of that enterprise. But I think we should be skeptical on an even larger scale, particularly of the division into disciplines that we now take for granted. In fact, we're always trying to make bridging disciplines to break down the fact that the disciplines are wrong in the first place. The academic disciplines were only separated in the late 19th century, when universities suddenly burgeoned, because everybody began staking out claims to their own little bit of reality. Sociological theory, for most of my youth when I was trying to learn it and finally got to be exasperated by it, was really nothing more than elaborate attempts to claim sociology as an autonomous subject, separate from psychology, biology, or anthropology. There's a great deal of arbitrariness to the disciplines. They've become hardened as territorialism steps in and egos get involved.

I've sometimes just had to ignore the disciplines. If I have a certain problem, I go where the problem leads me. If that means that I have to learn something new, I just damned well have to learn it, whether it's brain evolution or constitutional law. To end my diatribe against "The Gods of the Copybook Headings" as Kipling called them, my own experience has been that all the interesting problems occur in the interstices between disciplines. And that's why you have to have fascinating places growing up like the Santa Fe Institute—places which refuse to accept disciplinary boundaries and go after problems, like complexity and chaos. That's where the interesting work in the future is going to be done. Within 20 years, most of what's now being done in the universities will look like a sort of giant obituary for dead thought.

Tiger: I agree completely with all of that, except maybe for the Santa Fe Institute. I'm afraid it may exemplify the phenomenal skill, also seen in the university, for semantic obfuscation—that is, the real poverty of language and the willingness to distort by hiding, to become holier than thou by being more subclaused than thou. This is a feature of the university structure that has become quite devastating. Academic talk, or worse, academic writing, is immune to so many of the real issues of life itself that the people using these linguistic modes end up being unable to do their job.

I think if *Skeptic* wanted to be really skeptical, it could be skeptical of the language used in universities. One of the reasons I like the magazine is because of the verve of the language. It's open and filled with energy, rather than jargon.

Fox: Agreed.

Tiger: It becomes painfully obvious that these professional languages are disguises and codes, rather than forms of communication. That is one of the most sinister features of the contemporary university.

Fox: Lionel, you're sounding like a deconstructionist.

Tiger: I don't think I am. I'm making a distinction between the language and the thing described. Heaven knows there are interesting things to write about human behavior, but the way a lot of our colleagues write, it's impossible to understand what those behaviors are. There are endless arrays of linguistic protection against reality. That has been a disastrous, non-contribution of the university system.

Miele: I want to thank you both for taking the time to speak with *Skeptic* and hope we can all do it again 25 years from now—but I'm skeptical.

References

Ardrey, R. 1961. *African Genesis*. London: Collins.
___. 1966. The *Territorial Imperative*. New York: Atheneum.
___. 1970. The Social Contract. New York: Atheneum.
Argyros, A. 1991. *A Blessed Rage for Order: Deconstruction, Evolution, and Chaos*. Ann Arbor: University of Michigan Press.
Betzig, L. 1986. *Despotism and Differential Reproduction: A Darwinian View of History*. New York: Aldine.
Cavalli-Sforza, L., P. Menozzi, and A. Piazza. 1994. *The History and Geography of Human Genes*. Princeton University Press.
Cavalli-Sforza, L. and F. Cavalli-Sforza. 1995. *The Great Human Diasporas—The History of Diversity and Evolution*. Reading, Mass.: Addison-Wesley.
Carroll, J. 1995. "Evolution and Literary Theory." *Human Nature*, Vol. #6 No. #2, pp. 119-134.
D'Andrade, R. 1995. "Moral Models in Anthropology." *Current Anthropology*, June, Vol. 36, pp. 399-408.
Fox, R. 1995. "Sexual Conflict in the Epics." *Human Nature*, Vol. 6 #2, pp. 135-144.
Hartung, J.1995. "Love Thy Neighbor" *Skeptic* Vol. 3, No. 4, pp. 86-99.
Lemann, N. 1995. "The Structure of Success in America." Atlantic Monthly, August, Vol. 276, #2, pp. 41-60.
___. 1995. "The Great Sorting." Atlantic Monthly, September, Vol. 276, #3, pp. 84-100.
Morris, D. 1994. *The Human Animal: A Natural History of the Human Species*. BBC-TV/The Learning Channel/Discovery Production Video Series.
Morris, D. *The Human Animal: A Personal View of the Human Species*. New York: Crown Publishers.
Read, C. 1920. *The Origin of Man and His Superstitions*. Cambridge University Press.
Tiger, L. and R. Fox. 1971. *The Imperial Animal*. New York: Holt, Rinehart, and Winston.
___. 1989. The Imperial Animal. (2nd Edition, with Foreword by Konrad Lorenz and New Introduction by the Authors.) New York: Henry Holt.
Wilson, E. O. 1975. *Sociobiology: The New Synthesis*. Cambridge: Harvard University Press.
Wilson, J. Q. 1993. *The Moral Sense*. New York: The Free Press.

Frank Miele

CHAPTER 4

THE IONIAN INSTAURATION

E. O. WILSON ON HIS CONTROVERSIAL BOOK: *CONSILIENCE: THE UNITY OF KNOWLEDGE*

Edward Osborne Wilson has had more impact on the borderland between the natural and the social sciences than anyone alive, possibly anyone since Charles Darwin. Perhaps that is why so astute an observer of the interrelationships between science and culture in contemporary America as Tom Wolfe, best-selling author of *The Electric Kool-Aid Acid Test*, *Painted Word*, *The Right Stuff*, and *Bonfire of the Vanities*, has dubbed Wilson "Darwin II," the intellectual guide not to "the evolution of the species…but to the nature of our own precious inner selves" (Wolfe, 1996, 214). Pellegrino University Research Professor and Honorary Curator in Entomology of the Museum of Comparative Zoology at Harvard University, Wilson is a member of the National Academy of Sciences, recipient of the Swedish Academy's 1990 Crafoord Prize (the world's most distinguished environmental biology award), the International Prize for Biology from Japan (1993), and the world's leading expert on ants. He is also co-originator of the theory of island biogeography, a champion of preserving the rain forests and biodiversity (for which he has received the 1990 Gold Medal of the Worldwide Fund for Nature and the 1995 Audubon Medal of the National Audubon Society), and winner of two Pulitzer prizes in general non-fiction for *On Human Nature* (1978) and *The Ants* (1990), the latter along with co-author Bert Hölldobler.

Wilson's book, *Sociobiology: The New Synthesis* (1975), laid out the evidence for the evolutionary and genetic basis of social behavior across the animal kingdom. The fact that he dared to include a final chapter, "Man: From Sociobiology to Sociology," set off the Sociobiology Wars of the late 1970s. That conflict was fought not only in academic journals and at conferences, but in campus protests and demonstrations. At a special symposium on sociobiology at the 1978 meeting of the American Association for the Advancement of Science (AAAS), a member of the International Committee Against Racism (INCAR) dumped a pitcher of water on Wilson's head as she and fellow demonstrators chanted, "Wilson, you're all wet!" SKEPTIC Editorial Advisory Board member Stephen Jay Gould chastised the crowd by quoting Lenin on the inappropriateness of violence for mere radical posturing, as opposed to the attainment of worthy political goals, and referred to the incident, again in Lenin's words, as "an infantile disorder of socialism." SKEPTIC Editorial Advisory Board member

Napoleon Chagnon tried to make his way to the podium to eject the protesters (Wilson, 1994, 348-350).

While his views on human nature have been anathema to many self-identified leftists, Wilson's passionate pleas on behalf of preserving the rain forests and on the value of biodiversity have often been criticized by many on the political right. Julian Simon, the late pro-growth economist, praised Wilson as a "superb scientist," but questioned the estimates of the increasing rates of animal extinctions that Wilson has cited in defense of environmental preservation (see Chapter 16). And in one of his zanier zingers, even for him, conservative radio talk show host Rush Limbaugh denounced Wilson's article "Is Humanity Suicidal?" in the *New York Times Magazine* (1993), as the ranting of a "radical liberal environmentalist who teaches at Harvard," and who is "just the kind of guy the Clinton administration listens to and who is teaching our children" (Limbaugh, 1993). This must have come as news not only to Wilson and the Clinton administration, but all those left-wing demonstrators at Harvard.

The model of a kind and gentlemanly scholar, now nearing 70 at the time of this interview, Wilson has the ability both to ignite intellectual arguments that draw common fire from otherwise bitter enemies, and also to inspire affection and admiration from most of those who disagree vehemently with him, and all of those who know him personally. His book, *Consilience: The Unity of Knowledge* (Knopf, 1998), set off yet another conflagration in the intellectual and academic journals and conferences and their associated e-mail networks.

In *Consilience*, Wilson argues that the natural sciences, based as they are on extensions of Darwin's seminal vision, have now reached a state of sufficient maturity that they can light the way through the borderlands connecting those disciplines with not only the social sciences, but also religion, ethics, politics, and the arts and humanities. The idea of basing anything on evolution is anathema to some, and the biologizing of social science sets off an immediate protest by others. Not only religious and political zealots, but contemporary philosophers like Richard Rorty and Skeptic icons like the late Stephen Jay Gould, the late Carl Sagan, and Michael Shermer, question not only our ability, but the desirability of trying to unite such traditionally discrete domains as science and religion. To Wilson, eschewing consilience in favor of maintaining epistemological boundaries between the disciplines has produced a result analogous to fragmentation in a computer system, in which the scattering of files over non-contiguous areas of memory increasingly degrades performance until it eventually brings the system to its knees. In its place Wilson calls for an instauration, a term coined by Sir Francis Bacon (one of the heroes of *Consilience*), meaning "a new beginning" (1998, p. 27), of the dream of the pre-Socratic Greek philosophers, who walked the Ionian shores and searched for unity in the nature of all things. *Consilience* may well be the most influential book in setting an overall direction for contemporary intellectual thought since Karl Popper's 1943 *The Open Society and Its Enemies*. Here is E. O. Wilson on Consilience and its Discontents.

Figure 4.0: E. O. Wilson

Miele: In previous books (Wilson, 1975, 1984) you've given us the words "Sociobiology" and "Biophilia". "Consilience" is not your word. Whose word is it, what does it mean, and why do you choose it as your title?

Wilson: The word was introduced in 1840 by William Whewell in his master work, *The Philosophy of the Inductive Sciences*. He used it to mean the interlocking of cause and effect explanations across disciplines. Whewell thought consilience provided the key to understanding the natural sciences as a whole; to use his words, to "proving the validity of theory," by which he meant that theories which matched across disciplines, say chemistry and physics, or chemistry and biology, were consilient and therefore more likely to be true.

The word has been used sparingly since Whewell. I chose it because it has retained its purity, because of its rarity. The other words that might be used are "coherence" and "inter-connectedness," but these much more common terms have many meanings, only one of which is the same as consilience. I hope that I can now introduce the word "consilience" into the mainstream to capture one of the key qualities of the scientific method.

Miele: It struck me as odd that you'd choose a word from Whewell, who, I believe, was an opponent of Darwinism (Ruse, 132).

Wilson: If that's true, he didn't know consilience when he saw it!

Miele: You also refer to the "Ionian Enchantment," which you tell us physicist and historian Gerald Holton (1995, 4) defined as: "a belief in the unity of the sciences—a conviction, far deeper than a mere working proposition, that the world is orderly and can be explained by a small number of natural laws. Its roots [hence the name] go back to Thales of Miletus, in Ionia, in the sixth century BCE" (1998, p. 4).

I take this as an attempt on your part to get back to the founders of Western scientific philosophy—the pre-Socratic Greek philosophers. But elsewhere, you pay homage not only to your own upbringing as a Southern Baptist in Alabama, but also to the Biblical tradition which underlies Judaism, Christianity, and Islam, when you say that Chinese philosophy missed the scientific boat when it "abandoned the idea of a supreme being with personal and creative properties. No rational Author of Nature existed in their universe; consequently, the objects they meticulously described did not follow universal principles, but instead operated within particular rules followed by those entities in the cosmic order. In the absence of a compelling need for the notion of general laws—thoughts in the mind of God, so to speak—little or no search was made for them" (1998, 31).

Economist Kenneth Boulding (1984) has made a similar statement to the effect that by partitioning the world into the sacred and the profane, those traditions allowed for the growth of science, at least as it pertains to the profane. But isn't there an alternative interpretation that not only the Biblical tradition, but Socratic and post-Socratic philosophy, instead constitute a gigantic wrong turn away from the Ionian Enchantment and from which we are only now grudgingly regaining momentum, and then only in circumscribed areas?

Wilson: The explanation of slowing of Chinese science we owe to Joseph Needham (1978), who is the principal Western authority on the subject. The point you make is very important. It appears to me that the Judaeo-Christian tradition worked both ways. On the one hand, as you say, it created a dualistic view of the universe that impeded the thought of even the best of the Enlightenment philosophers. Notice, for example, how Descartes probably created dualism in order to avoid clashing with the prevailing Christian theology of his time and therefore getting in serious trouble. Note how Newton believed that there were, in fact, two books to be read—the Book of Nature and the Book of God's Word and that he actually spent time on both. He succeeded spectacularly on the first, and failed totally in the second.

On the other hand, there was the view that the world is orderly. In the minds of many of the pioneers of the scientific revolution, that reason was God. So the search for abstract laws seemed logical to them, which turned out to be a correct methodological assumption.

Miele: Isn't there in the Biblical and the Socratic traditions an attempt to understand the world by seeking to explain not how things work scientifically, but rather what they mean moralistically? Doesn't that temptation eventually lead us into deconstructionism?

Wilson: It certainly would if we were to ignore virtually everything we know about biology, including evolution. We now understand how organic systems, including our own brains and minds, were self-assembled, and how apparent purpose evolved as a property of the biological human brain. The latter was, of course, the triumph of Darwinian thought, which was perceived correctly well over a hundred years ago as the principal threat to the theological concept of a purposeful God.

Miele: You pay homage to the Enlightenment thinkers and tell readers that they basically got it right in their "vision of secular knowledge in the service of human rights and human progress," which you consider to be "the West's greatest contribution to civilization" (1998, 14). But wasn't the

69

Enlightenment yet another wrong turn, and one you have been very critical of, in its strong preference for environmentalist, nurturist explanations of human behavior, as opposed to hereditarian, naturist, or interactionist explanations? Shouldn't you have drawn a distinction between the French Enlightenment, which I would argue is the godfather of Marxism, and what Paul Hollander (1988) has dubbed "the Adversary Culture" and the Scottish Enlightenment of David Hume, which was godfather to Adam Smith and later Charles Darwin and Sir Francis Galton?

Wilson: I couldn't argue against that distinction. Whoever injected the virus of perfectibility, and I suppose that was indeed a French influence, certainly made the perversion of Enlightenment thought possible. I hope I spelled out in enough detail in *Consilience* the real risk of perfectibilist Enlightenment thought, thought that leads to the idea that there is one system of government, one method of systematic description that must rule and which can bring humanity to perfection. That's certainly one of the most dangerous ideas of the 20th century.

Miele: But isn't that exactly what your critics would say you're trying to do in *Consilience*, and even earlier in *Sociobiology*?

Wilson: I don't think I was ever accused of being perfectibilist.

Miele: No, but of having one viewpoint that everything gets folded into.

Wilson: If any critics try to put it that way, it would be a stretch. What I've done is to lay out not a system that could be translated into a perfect political or economic order, but quite the contrary, to provide a method for mapping the unknown with the aim of showing the connectedness across the principal domains of knowledge. I don't think those are the same thing.

Miele: Let me go back to Needham and Chinese philosophy. You point out how China was ahead of Europe in technology and science between the first and 13th centuries, but then fell behind because Chinese philosophy focused on "the harmonious, hierarchical relationships of entities, from stars down to mountains and flowers and sand. In this world view the entities of Nature are inseparable and perpetually changing, not discrete and constant as perceived by the Enlightenment thinkers. As a result the Chinese never hit upon the entry point of abstraction and break-apart analytic research attained by European science in the seventeenth century" (1998, 30-31). Isn't that exactly what the developing disciplines of chaos and complexity theory, the Santa Fe Institute, and "the butterfly flapping its wings" are all telling us we need to do so we can bring science into the next millennium? Or, in your opinion, is that stuff just so much nonsense?

Wilson: Oh no! First let me say that Chinese science made tremendous strides. It's just that they slowed down because of their failure to emphasize reduction and abstract law. I don't see the mathematical models and forms of reasoning associated with complexity and chaos theory as being Chinese at all in this sense. In fact, they are exquisitely Western and consilient in approach. For example, the best of chaos theory is an attempt to explain the extremely complex patterns in nature by a relatively small number of deterministic equations.

Miele: This brings me to the diagram on page 5 of *Sociobiology* in contrast to the ones on pages 9 and 10 of *Consilience* (See Figures 4.1, 4.2a and 4.2b.). Back in 1975 you were accused of preaching an amalgam of philosophical reductionism, genetic determinism, and an attempt at "disciplinary imperialism" (see Caplan, 1978; Takacs, 1996, and Wilson, 1994). Now in 1998, your discussing the unity of knowledge, including even art, literature, and religion. But I sense you've doing it from a perspective of, if not holism, at least emergence. A recent article in *Science* by Nigel Williams (1997) summarized the CIBA foundation symposium that provided a lengthy critique of reductionism, with protein folding being the principal case in point. So how has your view of reductionism versus holism changed? Is there a middle ground of emergence that allows us to explain things like protein folding? Or is emergence only a nice name we put on our ignorance?

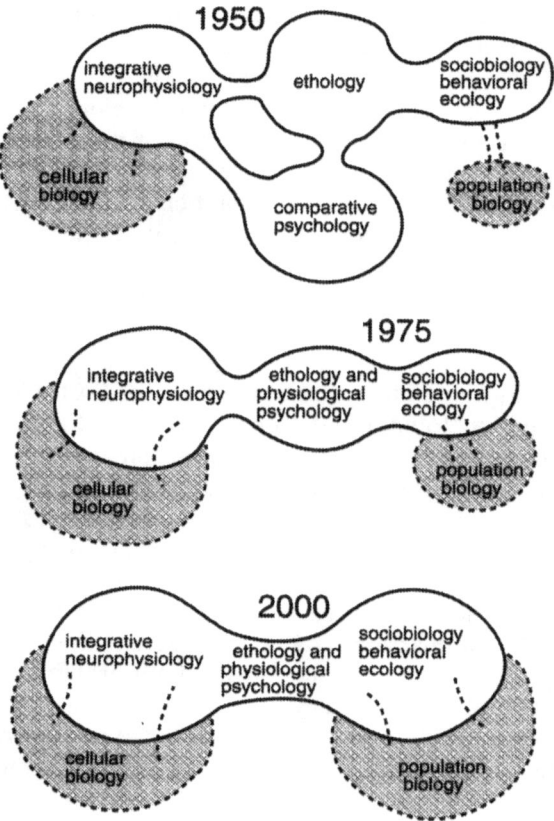

Figure 4.1: The conventional view of the separation of the disciplines, based on impermeable epistemological boundaries. (Adapted from page 9 in Wilson, 1998.)

Wilson: No, I think emergence is a name we put on patterns that cannot be easily explained by means of reconstruction from the elements revealed through reductionistic analysis. In theory, we should be able to predict many of the complex patterns at higher levels of organization, which we call emergent, from a knowledge of the constituent elements and their interactions. In most cases, especially when you get into biology, this would require massive computational capacity.

Unfortunately, emergence tends to come across as a mystical concept because going from a knowledge of the elements to predicting the complex interactive patterns is such a difficult step. It has been tempting for some thinkers to say that there are truly new phenomena that are independent of anything revealed by reductionistic analysis and that the connection between the constituent elements and the higher levels of organization cannot be made. I believe that when most scientists think it through, they would not accept that view. Scientists think all the time about complexity at whatever level they are investigating, the cell as a whole, or ecosystems as a whole. But at the same time, the most successful scientists proceed through reduction by analyzing that system into its constituent elements with the aim of reconstructing those elements back into an understanding of the complex whole.

As I pointed out in *Consilience*, the level of complexity of the re-synthesis from a pretty complete knowledge of the elements and how they interact up to protein synthesis is a slow process and it hasn't been achieved yet. When it is we will have taken both reduction and synthesis to emergent levels, all the way from quantum physics to protein synthesis. Going beyond that to an organism, an ecosystem, or a mind may prove ultimately to be beyond our computational capacity, but it's a worthy goal, the pursuit of which reflects the confidence we have gained in the consilient approach by its success in explaining less complex systems.

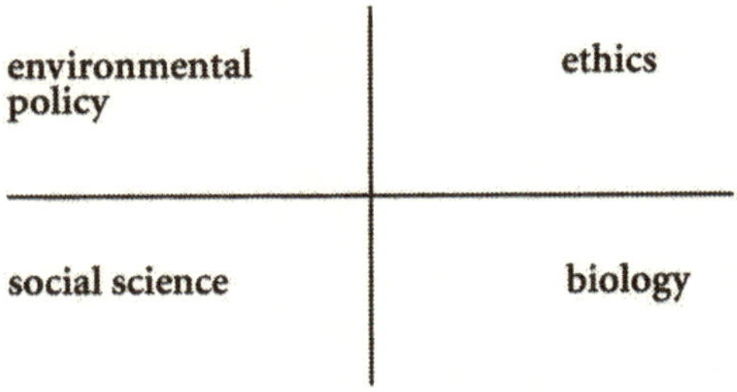

---

environmental policy	ethics
social science	biology

Figure 4.2a: Wilson's 1975 conception of the projected evolution of the behavioral and biological sciences from 1950 to the year 2000, with integrative neurophysiology and the new science of sociobiology absorbing the other disciplines. Some of Wilson's critics saw this as an attempt at "disciplinary imperialism." (Adapted from Figure 2-1, in Wilson, 1975.)

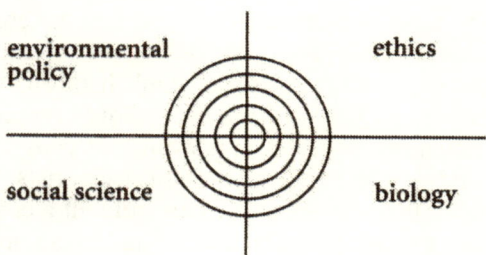

environmental policy

ethics

social science

biology

Figure 4.2b: Wilson's 1998 view of how the consilient approach can dissolve these epistemological boundaries, with a "kinder and gentler" sociobiology providing the focal point from which to unify what otherwise appear to be separate disciplines. (Adapted from page 10 in Wilson, 1998.)

Miele: My question remains whether your dedication to reductionism is methodological, that it's pragmatically the best way to go, or is it metaphysical, that eventually, with enough knowledge, everything can be encompassed within our understanding?

Wilson: Both. Methodologically, reductionism has proven spectacularly successful across a large part of science. Metaphysically, we are speaking again of the Ionian approach to the physical world.

Miele: On page 13 of *Consilience* you say that every college student, public intellectual, and political leader, should be able to answer the question, "What is the relationship between science and the humanities, and how is it important to human welfare?" This seems to be the very heart of what's at issue in *Consilience*, why philosopher Richard Rorty (1998) criticized your thesis in the *Wilson Quarterly*, why your keynote speech at the 1996 meeting of Human Behavior and Evolution Society (HBES) got a standing ovation, but then John Beckstrom of Northwestern University Law School felt moved to send a letter to fellow HBES members telling them you were flirting with the naturalistic fallacy. Another HBES member, psychologist Irwin Silverman (1993, 1998), has argued that the only defensible political role for the behavioral scientist is no role at all—science and politics must be treated as dichotomous and often antithetical domains. Well known skeptics like Stephen Jay Gould, the late Carl Sagan, and Michael Shermer have all argued for treating religion and science as, to use Gould's neologism, NOMA—Non-Overlapping Magisteria. Can we unify knowledge through the consilient approach or should we compartmentalize knowledge and lead our lives as if we were split-brained? Isn't that what these critics of consilience are asking us to do?

Wilson: A comment on the side, which you can print or not—Frank Miele you are well informed! That's a compliment I can't resist making.

Miele: I can't resist taking it!

Wilson: These strong, even emotional objections that you cite are what I call the secular intellectuals' white flag. It's very comfortable at this point, and believe me, politically correct in an almost all-encompassing sense, to say that scientists should be humble and partition off their thinking about the physical world and religious phenomena. I believe no such thing. I just find religion one of the more interesting physical phenomena that we should be studying more vigorously.

Miele: Then let me toss you an even hotter potato. Wouldn't even those who want us to treat science and politics as separate, non-overlapping realms have to acknowledge that every political philosophy, including their own, has at its core a theory of human nature, even if only to deny that there is such a thing?

Wilson: You're exactly correct about that. This is the point I tried to make in *Consilience* in order to quiet forever the arguments against sociobiology by pointing out that everyone has a theory of human nature. (I know I won't succeed, but I tried.) The so-called opposition to sociobiology—the argument that there is no human nature, which has been so widespread and emotionally argued—is actually a theory of human nature. It says that whether the human brain evolved or was put there by God, it was constructed in such a way that it is equally likely to absorb any form of information, belief system, or custom.

When you think about that for a while, it is highly unlikely. In fact, some investigators like Charles Lumsden and myself (1981, 1983) have shown that it is almost impossible if the brain originated by organic evolution. We now have enough evidence from sensory physiology and cognitive neuroscience to show why that view is untrue. There is a human nature, and among the varieties of theories of human nature, you must choose one. There is a system of what I call epigenetic rules, algorithms that predispose people to choose certain options as opposed to others. I find this feeling that political and religious belief are somehow insulated from explanation and from testing in terms of their etiology to be a curious product of late 20th-century thought that we will do better to get past.

Miele: Then let's think about trying to build a political or religious system on a scientific basis. One of the defining characteristics of science, maybe the defining characteristic, is that it deals in corrigible, tentative statements. Yet we have this desire to ground our morals, our laws, in something eternal—or at least something we like to think is eternal. Suppose Thomas Jefferson had written, "I'm not really sure about all this, but based on the best evidence we have at the present, we cannot reject the null hypothesis that as far as the essential qualities necessary for citizenship, the differences between individuals, at least healthy adults, are insufficient to support any recognition of them as a principle of law." Would anybody have picked up his gun and fought in the American Revolution?

Wilson: I rather doubt it. That would be a fine Declaration for MIT, I suppose. As I explain in Chapter 11 of *Consilience*, "Ethics and Religion," I believe we have to choose between Transcendentalism, which in this context means that ethics are rooted in precepts that exist outside of humanity and are then acquired by humanity, either by study or by divine instruction, and Empiricism, which means that ethics and religious belief are human acquisitions that arise substantially from the predispositions in human nature. This forces us to think about where our ethics come from and what their values are, and how deeply they have affected our emotional lives.

It's true that wars are best conducted by those who believe that they are guided by absolutes, especially divine absolutes. That view, however, when shared by two opposing armies, is the reason why we have been wading in blood for millennia. It would appear to me that humanity can move toward a humanistic ethic in which ethical beliefs and precepts are, if I may use the term, "sacralized" by a strong consensus, based on a modern informed study of the evolutionary process and its consequences. I think such an ethic can only arise from a more consilient view toward the humanities. If we can first sacralize our ethical precepts by general strong agreement and then valorize them with the symbols of communal unity and agreement, while keeping them subject to modification based on further study and discussion, we could build an ethical system that is just as strong as the ones validated by religion and one that is much more likely to be global, as opposed to the systems we have now which are mostly tribal.

Miele: Well, let's get down to specifics. Can we come up with an evolutionarily-correct position or a consilience correct position on issues like abortion, animal rights, immigration, population and the environment, eugenics, genetic engineering, ethnic conflict and genocide? If you can't give me specific answers, aren't we just talking to each other?

Wilson: Well, that's what people have been doing anyway, just talking to each other. Those who have the largest majority on their side and the greatest passion at the moment are the ones whose views prevail. I think it would be foolish to claim to have reached a powerful, enduring, and sacralized position on any of those issues based on what we know now. The prospect, however, is that with more knowledge and open discussion based on that knowledge, we could reach consensus. For example, on the issue of population and the environment, the facts about the risk of over-population and the accelerating destruction of the environment are so clear-cut now that if they were laid out before everyone, which we have not properly done up to the present time, the world would most likely move to a pretty solid consensus, even overriding the strong natalist views of some traditional religions. That's one issue I'm pretty confident about.

Let me cite another one to show you that the empiricist view of the origin of ethical precepts is not just reckless relativism. The issue of incestuous sex, and even reproduction, has come up repeatedly. There have even been a small minority of thinkers who have suggested the incest taboo is just the last of the outdated sexual inhibitions. Now we know that is not the case. We have excellent evidence of the genetically dangerous effects of inbreeding. We know pretty precisely, in fact, what those increased risks are. We also know there is an epigenetic rule in mental development, the Westermarck effect, which, under most circumstances, causes people to become sexually desensitized to those with whom they are reared and to avoid having incest. Furthermore, we know that this is not only a trait of human cognitive development, but also of every one of the primate species that have been studied to date with this subject in mind. Incest avoidance is a very deep and ancient propensity of primate cognitive development. That evidence should effectively eliminate the last vestiges of the belief that incestuous sex and reproduction are permissible. That's a fairly easy one, but nonetheless it illustrates the principle that empirically derived ethical precepts can gain consensus and become established.

Miele: Do you want to try any of the others, or are they still in the talking stage?

Wilson: The latter. These, of course, are the ones that have so many ambiguities and cross-cutting preferences and consequences that one can't resolve them at the present time. Part of the problem is that the traditional religious positions, some of which have got to be wrong because many of them are in opposition to each other, have made any such discussion virtually taboo.

Miele: Let me take one that just comes to mind—a powerful, middle-aged male is alleged to have had a sexual involvement with a young female in a subordinate position. Is that something that we should look at based on human evolutionary history and say, "Hey, that's just doin' what comes natural." Or should we regard it as an unallowable exploitation of an individual in a precarious position, which is precisely why we need laws to overcome our evolutionary predispositions, rather than to reinforce them?

Wilson: Both statements are correct. No one in their right mind would commit that form of the naturalistic fallacy, saying that anyone should feel free to do whatever they feel emotionally inclined to do. Human societies, like all primate and, in fact, all vertebrate societies, are based upon compromises and competitive conflicts. Therefore inequalities and even despotism are possible consequences. Since human morality is based on contract and consensus formation, these need validation as societies become more complex. We should be especially cautious in arriving at a moral consensus about the propensities evolution has fitted us with, and the one you just cited is certainly an example.

Miele: We mentioned the naturalistic fallacy, which goes back to David Hume. When I asked Robin Fox about it, he said, "Hume was actually making a kind of anthropological observation. What Hume said is, 'Isn't it interesting, when you listen to people making moral arguments, how frequently and easily they jump from "Is" to "Ought," without giving you the intervening reasons.' Hume himself was quite willing to jump from 'Is' to 'Ought'" (see Chapter 3).

Wilson: I think you can go from "Is" to "Ought," and we just reviewed an example of how.

Miele: Well, conservatives who are happy with gender-specific roles like to say either that they're part of human nature or that God made us that way, except when it comes to sexual orientation. Then they say that homosexuality is unnatural. But a lot of liberal and left-wing individuals who wouldn't touch human nature otherwise say that sexual orientation is inborn and can't be changed. Doesn't that show that there's something inside us that resonates to the idea of natural law like the sympathetic vibration of a piano string? Does the fact that we make that transition so frequently and easily, as Hume originally noted, show that humans are predisposed to seek and accept "natural" explanations?

Wilson: I believe it shows that people tend to look after their own selfish interests if they can rationalize them. That's primarily a primate trait. It's also a rather lazy and ignorant way of thinking.

Miele: Dostoevsky's Grand Inquisitor mused that, "If there's no God, anything goes." Your whole point in arguing for consilience seems to be clearly opposed to an "anything goes" philosophy for the reasons you spelled out in the example of incest. So don't we need concepts like God, a soul, and free will in order to practice morality and keep us from following our selfish interests?

Wilson: The Grand Inquisitor's statement actually could be turned around to say, "If there is a God, anything goes," because we've seen how in the evolution of human societies God has been invoked with great passion to justify every form of criminality, up to and including slavery and genocide. So much for the need for God in order to have morality.

Miele: How about free will?

Wilson: I spelled out how the free will issue can be handled in Chapter Six of *Consilience*: "The Mind." To very briefly summarize, free will exists insofar as the ambit of ordinary human thought reaches. Perhaps a super-computer simulating the action of virtually every molecular system within every neuron of a person's brain might be able to predict in advance how that person is going to act. But that is a theoretical construct so far beyond what is practical, that effective human action cannot be determined. If taken to that level, the issue of free will becomes a fruitless argument. At the common sense level of an individual being able to follow a wide range of intentions, however, free will certainly exists.

Miele: One of the pillars of sociobiology was explaining eusociality (true or higher social behavior in animal species) in terms of haplodiploid sex determination (i.e., where males derive from unfertilized [haploid] eggs; females from fertilized [diploid] eggs). That's an example of reductionism at its best. But not all haplodiploid species are eusocial. Termites are not. And some eusocial species are not haplodiploid, the naked mole rat being an example. So if haplodiploidy is neither a necessary nor a sufficient condition for the evolution of eusociality, is the pillar of sociobiology starting to shake a little?

Wilson: Oh no. Sociobiology would do very well without kin selection. Kin selection is just one of many models derived from evolutionary biology that have been used to explain the origins of social behavior. I don't want to sound glib about this. I put it that strongly to emphasize that what we're referring to is not some kind of over-arching top-down system of explaining social behavior like those that have prevailed during most of the 19th and 20th centuries, including Freudianism and other classic grand theories. Sociobiology employs a natural science approach to explain the workings and origins of social behavior.

I happen to think that kin selection is a very powerful and well substantiated process, not only in social insects, but also in human beings. It doesn't explain everything, but it's very rare that a single factor like collateral selection (the selection among groups other than parent-offspring) is all-explanatory. A single factor can introduce a major biasing in evolution. The concept of collateral selection has held up very well. In the case of the social *Hymenoptera* (the order of insects that includes ants, wasps, and bees), there is plenty of evidence for other biasing factors besides haplodiploidy, which fits all of the eusocial insects, except for the termites. Species that tend to nest close to one another in the same place, such as solitary bees or wasps, are more likely to evolve into advanced social forms. So are species that take care of their young. I call these factors part of the Eusocial Complex. Multiple factor models are very common in evolutionary biology and they are testable.

Miele: Well, if the score in the game between sociobiology and its critics was 1-to-1 back in 1975, what is the score now?

Wilson: When you recognize that the term Evolutionary Psychology is the same thing as Human Sociobiology, I would say it's now 10-1, in favor of Sociobiology.

Miele: Then why the need to change the name to Evolutionary Psychology? (For an introduction to the field see Chapter 1)

Wilson: This is something of a puzzle. Evolutionary Psychology is the same subject as Human Sociobiology, but it's been picked up because it contains the word "psychology." Most of the investigators in that area obtain employment in departments of psychology or anthropology. It would be harder to obtain one of those job slots by identifying yourself as a biologist of any kind.

Miele: Why is that? Political correctness?

Wilson: No. I was going to come to that in a moment. The reason is the classic division of academic departments. The psychology department wants to cover all fields of psychology. So it's a little easier to get a position in the psychology department if you bill yourself as some kind of psychologist, rather than some kind of biologist. However, since Evolutionary Psychology came along after the Sociobiology Wars of the 1970s, it's a little bit more politically correct.

Miele: Let's move on to even less politically correct topics. Suppose the comet Chicxulub failed to hit the earth, or that some of the great extinctions of the past had not taken place. Is there a consistent set of evolutionary pressures that would have led some species of dinosaur, or ant, or some other group, to develop a neurological complexity sufficient to support the cognitive complexity, behavioral complexity, even moral complexity of modern humans? If there's life elsewhere in the universe, are there universal evolutionary pressures that would lead to neurological, behavioral, and even moral complexity over time? Or is that 19th-century progressivism?

Wilson: At the present time, that is an unanswerable, but very good question. We may have an answer when the natural sciences have become strong and consilient enough, because the answer requires our first reconstructing human cognitive evolution and the circumstances under which it arose, and then seeing how probable that same set of conditions would be in other large non-terrestrial biospheres. At the present time, it is virtually science fiction speculation. I did discuss the issue of human cognitive evolution in *Promethean Fire*, a book I wrote with Charles Lumsden (Lumsden and Wilson, 1983) where we developed the first theory of gene-culture coevolution. We reviewed the circumstances that might have contributed to this one species of Old World primate starting the runaway process of gene-culture coevolution.

Gene-culture coevolution was very likely a runaway process. Reaching the threshold to trigger that process must be very difficult because, obviously, out of hundreds of millions of species that have existed throughout geologic time, only one crossed it. There have been innumerable hypotheses about why that one species crossed the threshold—bipedalism, the challenge of a drying environment, and so on. Some day we may know why, but until then I think it's useless to speculate about intelligent life on other planets.

Miele: What about the whole question of progress in evolution? Is there progress or are we just reading our own value system into the geological record?

Wilson: It's very unfashionable, especially with the dull postmodernist cast that has seized some of our popular science writers, to speak of progress. The difficulty in the word is easily solved by simple semantic distinction. Everyone would agree that progress, in the sense of moving toward some predestined ultimate high point, does not exist in organic evolution. If, however, you mean the appearance of ever more cognitively complex and environmentally controlling species as a result of natural selection working over the course of evolutionary time, then progress does exist.

Miele: You spoke of gene-culture coevolution and you and Lumsden used the term, "culturgen." Richard Dawkins came up with the term "meme," and that one seems to have stuck, where culturgen did not. Why the need for two terms?

Wilson: Dawkins used meme in his best seller *The Selfish Gene* (1976) before Lumsden and I came up with the term culturgen. True to the consilient approach, Lumsden and I wanted to identify our term with a unit of cognition and so we used the node linkage nomenclature of cognitive scientists in the early 1980s. We suggested that if these nodes could be defined precisely, neurophysiologically as well as semantically, they would prove to be the ultimate elements of culture. So we called them culturgens. I'm very happy to call them memes and recognize that whatever that unit is, it still remains elusive.

Miele: But doesn't your terminology emphasize that there's an underlying genetic biological mechanism?

Wilson: Yes, and that's why I would have preferred to have culturgen gain acceptance, but the world has taken to the word meme. I gladly accede to that and just hope that the word meme will eventually fit such a neurobiologically defined unit.

Miele: How far can we carry the concept of gene-culture coevolution? Let's imagine two human societies, one that is communal and cooperative like the Quakers or the Amish, and the other similar to, say, Mafia families. Over the course of time, isn't there going to be a biasing of both the culture and the genes in those two different groups for two different types of behavior?

Wilson: If they were totally reproductively closed, that is, breeding only among themselves, and this reproductive isolation was maintained by a great deal of selective pressure favoring certain predispositions as opposed to others, it could happen. But it would take quite a few generations. I'm not really sure that the behavioral predispositions of Mafia-type groups are that different from other groups. But let's assume that some hypothetical group did stress a different morality or set of behavioral rules which were maintained for enough time for the genes predisposing one to that type of behavior to be favored, you could have a divergence between groups. Lumsden and I calculated that under certain conditions you could expect to see a substitution of one gene favoring a certain type of personality over another gene in something like 50 generations, the order of a millennium. The point is that societies with sufficiently different modes of conduct from other societies don't hold together long enough to create that type of genetic diversity between groups.

Miele: But doesn't any invention, or a new religion with different rules for marriage, or some new hygienic practice, immediately create some form of evolutionary pressure that affects the genetic composition of the next generation?

Wilson: It could. I'm not just being cautious in order to be politically correct. I'm being cautious because I could easily make a foolish misstep here. In theory what you say is correct. But in reality there are so many other restraining conditions as to make the likelihood of a single cultural change producing a corresponding genetic change problematic. From the viewpoint of bringing consilience to the natural and social sciences, the subject of the genetic underpinnings of human nature has been skirted around and avoided so long that our knowledge is in a primitive state at present. It's risky, even apart from political correctness or incorrectness, for a scientist to speculate on the subject.

Miele: On page 53 of *Consilience* you write that the defining characteristics of science are: (1) repeatability, (2) economy (sometimes called parsimony or elegance), (3) mensuration (measurement), (4) heuristics (productivity, progress), and finally, (5) consilience. Clearly those are rules by which we can judge one scientific theory against another. But on what basis do we decide whether we want to play by that set of rules? Isn't that an initial value choice, rather than a scientific question, as Jacques Monod argued in *Chance and Necessity* (1971)? Or does Darwinian selection provide a means by which to judge theories and meta-theories by how well they work and survive?

Wilson: It's the latter, of course. It's what works. It's not whether we have some innate preference or ideological derivative preference for one way of doing science over another. *Consilience* provides a natural history of what has proved to work in terms of providing logically and pragmatically interconnected theories that are capable of successfully explaining a wide range of phenomena, and often in a predictive way.

Miele: At *Skeptic*, we deal with science, but we also deal with a lot of pseudoscience, pseudohistory, and downright charlatanism. I want you to give me your reaction to the Frank Miele test as to whether something is science or pseudoscience, history or pseudohistory, or art or pseudoart (i.e., propaganda)—Does the scientist, the historian, or the artist already know the answers to all the questions they pretend to ask? In preparing interview questions or writing introductory articles for *Skeptic* (as opposed to technical writing jobs where I'm just describing the workings of an engineered product), I'm always amazed how I never really know where the article will lead me. It often takes me in terribly different directions than I had initially anticipated. Each answer ends up generating more questions, so that you never reach finality.

Wilson: If you have an open mind that's true.

Miele: So, do you accept my distinction between the real and the pseudo?

Wilson: I do indeed. The writer Eudora Welty once put it very well when she said, "How do I know what I think until I've written it?" A very good writer, like a very good scientist, sets out with some preconceptions and perhaps a creative urge to put something together, to tell a story, to describe an event or a phenomenon. In the case of a writer, from the depth of their own mind through introspection and reconstruction of narrative; for the scientist, perhaps much more by actual examination of phenomena, or running tests and the like. Although both are highly motivated to produce a coherent story or pattern at the end, they are creating it as they go along. That's the mark of an open mind, the willingness to take in new information, even surprising and contradictory information, weaving it in and modifying the story as you go along. I like the Frank Miele test.

Miele: So what questions has *Consilience* raised in the mind of the man whom best-selling author Tom Wolfe (1996) has dubbed Darwin II, who has mapped the road to our understanding "the nature of our own precious inner selves?"

Wilson: You're touching on what I think is one of the main virtues of the consilient approach. Mind you, what I'm doing is not expressing some original new idea of philosophy or science, but rather I'm trying to articulate what I consider to be the core, the spirit, and the most successful methods of the natural sciences, which are now in a position to influence the social sciences and humanities in a profound way. One of the great virtues of this approach is that it creates a map. It's what I call gap analysis, borrowing a term from conservation biology where we analyze the distribution of organisms in parks and natural ecosystems to decide where reserves should best be placed. The consilient world view creates a map of what we know, what we don't know, and what are the most accessible large areas awaiting exploration. This amounts to a replacement of the epistemological boundaries that have separated the natural sciences from the social sciences and humanities with the concept of a large and mostly unexplored borderland. That borderland is wide open to the disciplines of cognitive neuroscience, human behavioral genetics, evolutionary biology, including sociobiology, and environmental sciences. On the social science side, there are biological anthropology and cognitive psychology. By viewing the natural sciences, the social sciences, and the humanities in terms of consilience we can put to rest the idea that science has come to an end, another popular obsession making the rounds these days.

Miele: Thank you for a most enlightening and consilient interview.

References

Boulding, K. 1984. "Toward an Evolutionary Theology" in Montagu, A. *Science and Creationism*. New York: Oxford.

Caplan, A. (ed.). 1978. *The Sociobiology Debate: Readings on Ethical and Scientific Issues*. New York: Harper.

Dawkins, R. 1976. *The Selfish Gene*. NY: Oxford University Press.

Gould, S. J. 1996. "Nonoverlapping Magisteria." *Natural History*. March: 16-22, 60-62.

Hollander, P. 1988. *The Survival of the Adversary Culture*. New Brunswick, NJ: Transaction Books.

Hölldobler, B. and E. O. Wilson. 1990. *The Ants*. Cambridge, MA: Harvard University Press.

Holton, G., 1995. *Einstein, History, and Other Passions*. Woodbury, NY: American Institute of Physics Press.

Limbaugh, R. 1993. "Commentary on E. O. Wilson's 'Is Humanity Suicidal?'". Rush Limbaugh Radio Broadcast of 31 May.

Lumsden, C. and E. O. Wilson. 1981. *Genes, Mind, and Culture: the Coevolutionary Process*. Cambridge, MA: Harvard University Press.

____. 1983. *Promethean Fire: Reflections on the Origin of Mind*. Cambridge, MA: Harvard University Press.

Monod, J. 1971. *Chance and Necessity: An Essay on the Natural Philosophy of Modern Biology*. New York: Knopf.

Needham, J. 1978. *The Shorter Science and Civilisation in China*. (C.A. Rowan, ed.) New York: Cambridge University Press.

Popper, K. 1966 (1943). *The Open Society and Its Enemies*. Princeton, NJ: Princeton University Press.

Rorty, R. 1998. "Against Unity." *The Wilson Quarterly*. Winter: 28-38.

Ruse, M. 1996. *From Monads to Man: The Concept of Progress in Evolution*. Cambridge, MA: Harvard University Press.

Sagan, C. 1996. *Demon Haunted World*. New York: Random House.

Shermer, M. 1997. *Why People Believe Weird Things*. New York: W. H. Freeman

Silverman, I. 1993. "Sociobiology and Sociopolitics." *Address to the European Sociobiological Society*, August.

____. 1998. "Can Behavioral Science Change Society? Should We Want to Try?" *Biopolitics* (in press).

Takacs, D. 1996. *The Idea of Biodiversity: Philosophies of Paradise*. Baltimore, MD: Johns Hopkins University Press.

Williams, N. 1997. "Biologists Cut Reductionist Approach Down to Size." *Science* Vol. 277, #5325, July 25: 476-477.

Wilson, E. O. 1975. *Sociobiology: The New Synthesis*. Washington, DC: Island Press.

___. 1984. *Biophilia*. Cambridge, MA: Harvard University Press.

___. 1993. "Is Humanity Suicidal?" *New York Times Magazine*. 30 May: 24-29.

___. 1994. *Naturalist*. Cambridge, MA: Belknap Press.

___. 1998. *Consilience: The Unity of Knowledge*. New York: Knopf.

Wolfe, T. 1996. "Sorry, but Your Soul Just Died." *Forbes*. 2 December: 211-223.

Frank Miele

CHAPTER 5

DARWINISM— NEVER TOO OLD TO ROCK 'N' ROLL!:

SEX, VIOLENCE, AND OUR NEW UNDERSTANDING OF THE ROOTS OF COOPERATION AND CONFLICT

REVIEWS OF:

The Origins of Virtue: Human Instincts and the Evolution of Cooperation. By Matt Ridley. 1997. Viking. New York, NY. 295 pp. Hardbound, $24.95, ISBN 0-670-87449-3.

Good Natured: The Origins of Right and Wrong in Humans and Other Animals. By Frans de Waal. 1996. Harvard University Press. Cambridge, MA. 296 pp. Hardbound, $24.95, ISBN 0-674-35660-8.

Bonobo: the Forgotten Ape. By Frans de Waal & Frans Lanting. 1997. University of California Press. Berkeley, CA. 210 pp., 75 color plates, 9 black and white photographs, 9 maps and drawings. Hardbound, $39.95, ISBN 0-520-20535-9.

Demonic Males: Apes and the Origins of Human Violence. By Richard Wrangham and Dale Peterson. 1996. Houghton Mifflin. Boston, MA. 350 pp. Hardbound, $24.95, ISBN 0-395-69001-3.

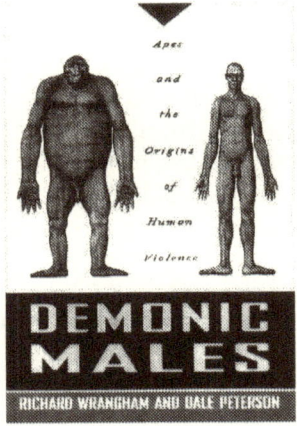

It all started in the 1960s when the way we see ourselves, our relationships, and the world around changed forever. Not the sex, drugs, and rock 'n' roll revolution, but the Darwinian Counter-Revolution—sex, violence, and a new understanding of cooperation and conflict! Since 1960, as Harvard anthropologist Richard Wrangham and writer Dale Peterson, authors of *Demonic Males*, explain, there has been "an enormous increase in the number of people who watched, rather than guessed, how animals live in the wild" and the result is that we now "know of animals manipulating, deceiving, attacking, and challenging each other in ways that were barely hinted at before" (156). We now have indisputable evidence of homicide, infanticide, and even genocide in cultures and species other than our own.

If the end of World War II saw a reaction against the extreme, simplistic, and nasty biological determinism of the turn of the century, the last 30+ years have witnessed a Darwinian counter-reaction to the hyper-Skinnerian restriction against assigning any role to consciousness, even among humans, and the hyper-Boasian *Weltanschauung* that excluded any reference to evolution or genetics in explaining our behavior, decreeing "much of what we ascribe to human nature is no more than a reaction to the restraints put upon us by our civilization" (quoted on p. 102).

A host of beautifully photographed nature documentaries has brought the best-known observational studies—George Schaller's work on lions, Jane Goodall's on chimpanzees, and Dian Fossey's on mountain gorillas—to the attention of even the general public. The intellectual impact of these megahours of observation, however, would have been minimal without the theoretical revolution in evolutionary thought brought about by William Hamilton's kin selection theory and Robert Trivers' theory of reciprocal altruism. Taken together, they allow us to build testable hypotheses which in turn allow us to answer many of the criticisms of those who would dismiss all such work as 'Just So Stories' (for example, Stephen Jay Gould's "The Darwinian Fundamentalists" in *The New York Review of Books*, June 12 and 26, 1997). For the last four decades skepticism has been stuck on the horns of a dilemma—whether 'To Keep On Keepin' On' against the earlier biological determinism or to 'Take a Walk on the Wild Side' by embracing behavioral Darwinism and skeptically scrutinizing the more recent socioeconomic cultural determinism. But when Barbara Ehrenreich and Janet McIntosh in an article entitled "The New Creationism: Biology Under Attack" (in the left-of-the-currently-very-centrist-center, *The Nation*, on June 9, 1997) approvingly quote no less than Stephen Jay Gould himself that "some facts and theories are truly universal (and true)—and no variety of cultural traditions can change that," you don't have to be wired to The Weather Channel to know that 'the times they are a-changin'' and that 'what goes around has indeed finally come around.'

As its title suggests, Wrangham's *Demonic Males* is the most rad of these books. Marshaling the anthropological and primatological data, Wrangham shows that "the system of communities defended by related men is a human universal that crosses space and time, so established a pattern that even writers of science fiction rarely think to challenge it" (25). It seems that for both chimpanzee and human males it isn't just Saturday night that's 'all right for fightin'.' We share with the chimps, "but with no other species, a uniquely violent pattern of lethal intergroup aggression visited by males on neighboring communities" (26) and a "system of intense, male-initiated territorial aggression, including lethal raiding into neighboring communities in search of vulnerable enemies to attack and kill" (24). Like the chimps, humans live in "patrilineal, male-bonded communities where females routinely reduce the risks of inbreeding by moving to neighboring groups to mate" (24).

'Can I Have a Witness?' Jane Goodall found that "about 30 percent of adult male chimpanzees died from aggression—about the same percentage Chagnon found in the Yanomamo" (70). A global assessment of "thirty-one hunter-gatherer societies found that 64 percent of them engaged in warfare once every two years, 26 percent fought wars less often, and only 10 percent were considered to fight wars rarely or never" (75). Crime statistics "from hunter-gatherer communities, tribal societies, and medieval and modern nation-states all uncover the same fundamental pattern....the probability that a same-sex murder has been committed by a man, not a woman, ranges from 92 percent to 100 percent" (115). A survey of 93 societies around the world found that men "hold the bulk of political power in all of them" and "retain all important political positions for 88 percent of them. Outside the more formalized and public political systems, men in 84 percent of the societies held all the significant leadership roles in kin groups as well" (119).

If Wrangham shows that for human and chimpanzee males, the answer to the question, 'War? What is it good for?' is 'Relatively everything,' Frans de Waal's *Good Natured* shows 'We can work it out,' while his *Bonobo: The Forgotten Ape* offers the reassurance that at least the nearest relative of humans and chimps (previously called the pygmy chimp) does in fact, 'Make Love, Not War.' *Good Natured* includes 64 pages of photos that are not fluff, but photo essays on "Closeness," "Cognition and Empathy," "Help from a Friend," and "War and Peace," that provide prima facie evidence supporting de Waal's case. *Bonobo* contains plenty of important information, but is worth buying simply for the 75 color plates of co-author and internationally acclaimed nature photographer Frans Lanting.

An ethologist at the Yerkes Regional Primate Center in Atlanta, de Waal reprises the arguments in his earlier *Chimpanzee Politics: Power and Sex Among Apes* (Harper & Row, 1982) and *Peacemaking Among the Primates* (Harvard University Press, 1989), remixed with information from a back-up chorus of other primate and animal studies. De Waal tells us that rather than "human nature's being either fundamentally brutish or fundamentally noble, it is both—a more complex picture perhaps, but an infinitely more inspiring one" (5). We now know that chimpanzees display "a range of humanlike expressions, unmatched by any other nonhuman primate, to seek contact and reassurance" and that they "pout, whimper, yelp, beg with outstretched hand, or shake both hands so that the other will hurry and provide the calming contact so urgently needed" (61). They teach their young not only how to behave, but how they "ought to behave," what would be "viewed in moral terms if seen in humans" (60).

'What's goin' down?' Furtive matings frequently take place between receptive female chimpanzees and lower-ranking males (the proverbial 'sneaky fuckers' as they have become known in the field). De Waal observed that the low-ranking Dandy and a female "were courting each other surreptitiously. Dandy began to make advances to the female, at the same time restlessly looking around to see if any of the other males were watching. Male chimpanzees start their advances by sitting with their legs wide apart and revealing their erection. Precisely at the point when Dandy was exhibiting his sexual urge in this way, Luit, one of the older males, unexpectedly came around the corner. Dandy immediately dropped his hands over his penis, concealing it from view" (77).

Aren't all such anthropomorphic inferences hopelessly flawed? Hasn't rigorous skepticism taught students of animal behavior that no matter how compelling at first glance it is "wrong and naive to speak of animals wanting, intending, feeling, thinking, or expecting" and that "animals just behave; that is all we know, and all we will ever know about them?" De Waal responds to these rhetorical questions that it is the "use of anthropomorphism as a means to get at the truth, rather than as an end in itself, that sets its use in science apart from use by the layperson. The ultimate goal of the scientist is emphatically not to arrive at the most satisfactory projection of human feelings onto the animal, but rather at testable ideas and replicable observations" (62). The proper scientific question, de Waal explains, is not whether any creatures other than ourselves can experience emotions, but "which elements…are recognizable in other animals" (79).

Doesn't even this limited use of observation violate one of the cardinal rules of critical thinking, the principle of parsimony—not to invoke higher capacities if the phenomenon can be explained by lower ones? De Waal rightly points out that evolution produces a parsimony of its own, so that "if closely related species act the same, the underlying process probably is the same too. The alternative would be to assume the evolution of divergent processes for similar behavior; a highly uneconomic assumption for organisms with only a few million years of separate evolution. If we do not normally propose different causes for the same behavior in, say, tigers and lions, there is no good reason to do so for humans and chimpanzees, which are genetically as close or closer" (64). Unless, of course, when dealing with humans one adopts what Vincent Sarich has dubbed 'behavioral creationism,' the belief that human behavior is somehow special and exempt from the laws of genetics and evolution. In which case one is guilty of a multiple-count violation of the principle of parsimony by first arguing that cultural evolution has undone the similarities that initially existed between human behavior and that of other primates and then reproduced that same behavior in all the cultures of the world.

According to de Waal, for humans and other primates, 'Break up to make up, is what we do.' He proposes a relational model that views "aggressive behavior as resulting from conflicts of interest between individuals who share a history (and a future). It assumes an equilibrium between tendencies that pull individuals apart and those that bring them together. It focuses on individuals drawn together by attachment and a sense of belonging to the same group" (173). De Waal further addresses the criticism, as in the *Seville Statement on Violence*, that Darwinian reasoning serves to justify aggression and violence. He counters that Darwinians are not "asking anyone to admire or encourage the behavior, only to step back enough to see that it is part and parcel of the social dynamics around us. So much so, in fact, that we may seriously question the tenet that aggressive behavior is by its very nature antisocial" (183). Dismissing absolute peace as utopian, de Waal concludes that there are "only two realistic alternatives in an imperfect world of limited resources: (1) unmitigated competition, or (2) a social order partly shaped and upheld by violence" and that monkeys, apes, humans, and a host of others opted for making "the use of force part of the solution to the use of force" (183) and that through "these regulatory functions an order is created that in complexity far exceeds that of animals, such as grazing herds of bovids, characterized by low levels of competition owing to evenly distributed resources" (184).

De Waal's message is mostly upbeat. Human biocultural evolution has proceeded because "our ancestors began to learn how to preserve peace and order—hence how to keep their group united against external threats—without sacrificing legitimate individual interests. They came to judge behavior that systematically undermined the social fabric as 'wrong' and behavior that made a community worthwhile to live in as 'right.' Increasingly, they began to keep an eye on each other to make sure that their society functioned in the way they wanted it to function. Conscious community concern is at the heart of human morality" (207-208).

Matt Ridley's *The Origins of Virtue* picks up that beat to explain how Trivers' concept of reciprocal altruism allowed the mind to evolve a capacity for calculating the costs and benefits of social exchange. The beat goes on as Ridley, former American editor of *The Economist*, and author of *The Red Queen: Sex and the Evolution of Human Nature* (MacMillan, 1993), uses game theory to show how this propensity eventually allowed our ancestors to calculate the costs and benefits of goods and services and build larger societies with complex economies. He notes that in "the two cleverest families of land-dwelling mammals—the primates and the carnivores—there is a tight correlation between brain size and the social group. The bigger the society in which the individual lives, the bigger its neocortex relative to the rest of the brain. To thrive in a complex society, you need a big brain. To acquire a big brain, you need to live in a complex society. Whichever way the logic goes, the correlation is compelling" (69).

Ridley then points out that "the human brain is not just better than that of other animals, it is different. And it is different in a fascinating way: it is equipped with special faculties to enable it to exploit reciprocity, to trade favours and to reap the benefits of social living" (131).

In Ridley's view sociality isn't just a matter of cold calculation, with the emotions being merely a blast from our evolutionary past. Rather, what Adam Smith termed the 'moral sentiments' are 'golden oldies, essential classics.' They are "problem-solving devices designed to make highly social creatures effective at using social relations to their genes' long-term advantage. They are a way of settling the conflict between short-term expediency and long-term prudence in favour of the latter" (133). Our emotions "alter the rewards of commitment problems, bringing forward to the present distant costs that would not have arisen in the rational calculation" (135). Ridley cites Antonio Damasio's study of patients with damage to their prefrontal lobes (*Descartes' Error: Emotion, Reason, and the Human Brain*, Avon, 1995). They have no loss of motor or sensory ability and do perfectly well on tests of memory or intelligence, but their lives are a disaster. Why? Because they "become so cold-blooded about rationally weighing all the facts before them that they cannot make up their minds" (144). Rather than the absence of emotions producing some Mr. Spock-like nirvana, there 'ain't no sunshine when they're gone.'

In passing, Ridley, who writes the "Down to Earth" column in the *London Sunday Telegraph*, debunks two elements of what Michael Shermer has termed the Beautiful People Myth—that humankind once lived in perfect harmony with nature before technology, phallocentrism, the Aryan invaders, the division of labor, capitalism, the Judaeo-Christian ethic, or Christopher Columbus destroyed them (*Skeptic* Vol. 5, No. 1). The first concerns Chief Seattle of the Duwamish Indians who, it is said, responded to the President Franklin Pierce's offer to buy the land, on which his people lived, "How can you buy or sell the sky? The land? ... This we know: the earth does not belong to man, man belongs to the earth." The Chief laid down such an ecologically conscious rap it's been covered by just about everybody in the business, including Al Gore. The problem is, the Chief never 'talked the talk, let alone walked the walk.' "The only report, made thirty years later, was that he praised the generosity of the great white chief in buying his land. The entire 'speech' is a work of modern fiction. It was written for an ABC television drama by a screenwriter and professor of film, Ted Perry, in 1971." What we do know about Chief Seattle is that "he was a slave owner and had killed almost all his enemies" (213-214). The second concerns the Kayapo people to whom the Brazilian government granted a reserve of 20,000 square miles, and to which the rock star, Sting contributed $2 million, so that they could protect and preserve their native, natural environment against the encroachments of the global economy. Within a few years of taking legal possession the Kayapo began "an enthusiastic programme of selling concessions to gold miners and loggers" (214).

The Battlegrounds of Bio-Science

Ridley concludes that the evolutionary message isn't that we're 'living on the eve of destruction' but that we should 'let the good times roll.' Since "the roots of social order are in our heads," and "we possess the instinctive capacities for creating not a perfectly harmonious and virtuous society, but a better one than we have at present," our task should be "to build our institutions in such a way that they draw out those instincts" (264).

All of four of these books are Darwinian 'solid gold' and belong on every skeptic's bookshelf. 'Roll Over Franz Boas, and Tell B.F. Skinner the news—Darwin Rules!'

Frank Miele

Part II

The Human Origins and Diversity Debates

Frank Miele

CHAPTER 6

THE SHADOW OF CALIBAN

AN INTRODUCTION TO THE TEMPESTUOUS HISTORY OF ANTHROPOLOGY

Figure 6.0: Chart of Caliban's Island — A Mental Map of Humanity's Attempts to Understand Human Nature.

In 1607 a sailor named Andrew Battell returned to England from imprisonment in Portuguese Africa, bringing with him tales of new worlds and half-man, half-beast monsters we now know to be chimpanzees and gorillas. (Peterson and Goodall, 2000). Then, in 1609, a fleet of ships set out from England to take Sir Thomas Gates, the new governor, to the struggling Jamestown colony in Virginia. The fleet was caught in a tempest off the coast of Bermuda. All the ships survived the storm, except the one carrying the governor. A year later, much to the surprise of the colonists, Sir Thomas and his crew landed in Jamestown. They had survived the storm, beached in Bermuda, and lived there for a year. By 1611, stories of the governor's miraculous return and of the Virginia colony appeared in church sermons, official government documents, and personal letters. They were filled with unflattering descriptions of the Indians, on whose "barbarous disposition, fair and noble treatment" had little effect. (Wright, 1964, pp. 88 - 89).

William Shakespeare drew upon these and similar accounts when he wrote *The Tempest* between 1610 and 1611. The play tells how Prospero, Duke of Milan, was deposed, driven into exile, and took refuge on a tropical island. Before he and his daughter Miranda arrived, the island was inhabited only by spirits and the half-man, half-beast monster, Caliban. Using the arcane knowledge in the books he brought with him, Prospero rules over the island and makes Caliban his slave. Rationalizing as much as lamenting, he describes Caliban as:

> A devil, a born devil,
> on whose *nature*
> *Nurture* can never stick;
> on whom my pains,
> Humanely taken, all,
> all lost, quite lost;
> —Act 4. Scene 1, 211-213 (emphasis added)

Some 250 years later, Charles Darwin's cousin, Sir Francis Galton, picked up the Bard's "alliterative antithesis" and bequeathed to modern behavioral science these terms derived from the Renaissance desire to throw off religious superstition, explore and conquer new worlds, and study the nature and nurture of human nature—along with the many tempestuous debates that have raged around that issue.

Since its inception anthropology has fallen under the Shadow of Caliban, that "born devil, on whose nature nurture can never stick." Some fear the concept of "human nature" is just political propaganda masquerading as science. For if we accept Prospero's verdict, social or political change is doomed to fail, or at best prove exceedingly difficult. If the genes keep culture "on a leash," (Lumsden and Wilson, 1981, p. 13) human differences are just the way Nature, if not God, made us; our nasty, brutish behavior is "just doin' what comes natural."

If, on the other hand, Prospero's words are just the rhetoric of the rulers and culture is primary, then "the modern world is this way because we, not God or Darwin, have made it this way." (Schwartz, 1987, p. 215). If we are able to recognize the shackles that tradition has laid upon us, we will be free to break them. That doesn't mean we can achieve utopia or that the task will be easy, but we can with confidence "take arms against a sea of oppressive 'isms,'—racism and sexism above all—and by opposing end them." Instead of an allegedly objective, scientific attempt to discover evolutionary or genetic "leashes," the job of anthropology should be to expose how and why such claims are inherently dangerous and inhumane.

In tracking the history of anthropological controversies it may be helpful to keep a chart of Caliban's Island as a mental map of humanity's attempts to understand itself. (Figure 1.) Rather than latitude, longitude, and elevation, Caliban's Island is mapped by the axes of Nurture versus Nature (left to right), Progressivism v. Primitivism (Up and Down), and Universalism v. Relativism (front to back).

The volcanic island rises from the surrounding sea of ignorance. The left side of the island depicts attempts to explain human behavior and differences in terms of cultural and environmental factors; the right side genetic and evolutionary explanations. Seen from the front, the island slopes upward from each end, showing human history as a progression from simpler to more complex. (The explanation for progress can be either cultural or genetic). On its back side, however, the island appears as a sheer cliff rising from the tiny shore of primitivism, lapped by the comforting waters of the sea of ignorance. At the summit of Caliban's Island is the hollowed-out core of a volcano. The eruptions of the volcano represent the increasing scientific understanding of human behavior in terms of universal principles, whether cultural, genetic, or both. The core of the now dead volcano depicts a precipitous plunge back into the abyss of ignorance.

SKULL AND BONES SOCIETIES

While Aristotle was the first to use the term "anthropology," it wasn't until the 19th century that the study of "the whole of man in all his various aspects" developed alongside other sciences (Coon, and Hunt, 1963, p. 133). Johann Friedrich Blumenbach is usually credited with being the founder of modern anthropology. In *De generis humani varietate nativa* (1776) he grouped humans into five major racial groups and argued they descended from a common human ancestor. Anthropology then developed largely as the science of describing and measuring different human groups (anthropometry), racially classifying them, and assigning each race its place in history. (For quite different evaluations of the scientific merit of early anthropology, compare Gould, 1996 [critical] against Baker, 1974 [sympathetic]).

The most widely used measure for racial classification was the cephalic index, introduced by Anders O. Retzius in 1840. It is defined as:

$$\text{Cephalic Index} = \frac{\text{Maximum Head Breadth} \times 100}{\text{Maximum Head Length}}$$

Based on the cephalic index, living individuals or groups were assigned to the categories of dolichocephalic (long-headed, CI < 75), mesocephalic (medium-headed, CI 75-79.99), or brachycephalic (broad-headed, CI > 80). For skulls of the deceased, the corresponding terms are cranial index, dolichocranic, mesocranic, and brachycranic. The cephalic index was then combined with the facial index (long-faced or short-faced), other anthropometric indices, and, for the living, hair color and form, and skin pigmentation to classify the individual (or specimen) according to race. Examples of such classification schemes from the heyday of anthropology are the familiar division of mankind into the Caucasoid, Mongoloid, and Negroid races, or of the Caucasoid race into the Nordic, Alpine, and Mediterranean subraces.

As fossil skulls, starting with the Neanderthal skull fragment in 1856, and cultural artifacts were also discovered, anthropology's scope expanded. In 1786, Sir William Jones demonstrated that Sanskrit, the language of the Hindu sacred texts, belonged to a linguistic super-family that included Ancient Greek, Latin, and most of the languages of Europe. His method of comparing languages for common elements of grammar and vocabulary was extended by Franz Bopp (1816) and Friedrich von Schlegel (1849). Language then joined skull shape, culture, and even mental traits, as would-be markers of race.

Early anthropology was an exercise in typology. Race was considered to be the primary Platonic essence, or type; culture (tools, pottery, and customs), language, and sometimes mental traits, were regarded as the manifestations of each type. This approach allowed vast amounts of data to be gathered and pigeon-holed. However, it lacked both an empirically testable theory of race and a method of validating its results. Without them, early anthropology relied upon speculations about migrations, conquests, hybridization, and degeneration in its alternative to the biblical account of human origins. Carleton Coon's 1939 description of the Corded Ware people in Figure 6.1, from his classic text *The Races of Europe*, shows how far this methodology had gone; and this was not that long ago.

Carleton Coon described the Corded Ware people by saying: "The length of the vault is great, well over 190 mm. in most instances; its breadth is slight, yielding the low mean cranial index of 71 and the height is great,

105

considerably exceeding the breadth. ...it is likely the Corded people came from somewhere north or east of the Black Sea." (Coon, 1939, pp. 107 - 108) Coon's caption for the photos shown in Figure 6-1a and 6-1b reads:

> A Nordic Dane of Jutish parentage [Figure 6.1a] who also shows Corded predominance. His face is of extreme length, a trait common among ancient Corded crania. This individual is the son of the classic Borreby man shown on Plate 5, Fig. 1 [Figure 6.1b]; *this is graphic evidence of the fact that ancient racial types may be repeated in toto in individuals of mixed ancestry. Only through the agency of such segregation is it possible to present this collection of basic European racial photographs.* [emphasis added]. (Coon, 1939, Plate 5, Figure 1; and Plate 27, Figure 3).

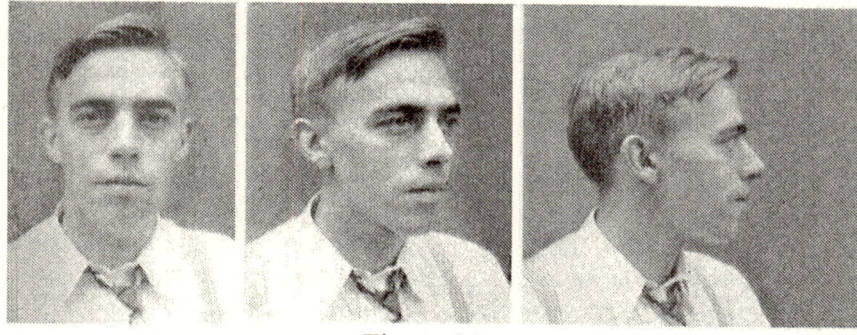

Figure 6.1a

Figure 6.1b

It required a synthesis of Darwin's theory of evolution by natural selection and Mendel's laws of inheritance—the synthetic theory of evolution worked out between 1936-1947 by Dobzhansky, Huxley, Mayr, Simpson, Rensch, and Stebbins—to provide a coherent model for the study of human origins and variation. (Mayr, 1982) With World War II intervening, however, the synthetic theory did not penetrate anthropological thinking until the 1950s. (Washburn, 1950; 1951) Attempts to fit the various racial classification schemes to the synthetic theory proved forced, if not impossible. One problem was that most anthropometric traits (e.g., head shape) depend on a number of genes. While clearly inherited, head shape does not follow the rules of simple Mendelian inheritance. Further, early genetic theory assumed that the anthropometric traits, no matter how many genes each one depended on, were inherited independently. Given that, why spend so much time taking all those different measurements of skulls? Increasingly many began to question the utility, and eventually the reality, of the concept of race. (Coon, Montagu, et al., 1950)

PROGRESSIVISM V. PRIMITIVISM

Charles Darwin changed forever the way we look at ourselves. Rather than "made in the image and likeness of God," Darwin argued that man shared a common origin with other creatures, especially apes. The explanation for human nature and human differences, between both individuals and groups, was therefore to be found by studying what we shared with other species.

Key to Darwin's theory was the underlying variation within the species. Nonetheless, the subtitle of *The Origin of Species*, *The Preservation of Favored Races in the Struggle for Life*, is much less frequently cited these days. In *The Descent of Man* (1874), the sage of Down House let it all hang out when he wrote, "It is not my intention here to describe the several so-called races of men; but I am about to inquire *what is the value of the differences between them* under a classificatory point of view *and how they have originated.*" (Darwin, 1874, emphasis added)

In 1859, the same year as the publication of Darwin's *Origin*, Paul Broca, best known for discovering the special role played by the temporal area of the dominant side of the brain in speech, founded the Anthropological Society of Paris. This was followed by the founding of the Anthropological Society of London in 1863, the Berlin Society for Anthropology, Ethnology, and Prehistory in 1869, the Anthropological Society of Vienna in 1870, the American Anthropological Association, and the American Association of Physical Anthropologists in 1902.

Between 1859 and 1945 the evolutionary perspective was the central principle of anthropology. Intertwined with Darwin's theory of biological evolution, however, was a theory of "progress," or cultural evolution. Herbert Spencer was the principal proponent of "progessivism," but the doctrine can be traced all the way back to the Greeks, as can its opposite, "primitivism." (For a detailed discussion of progressivism and primitivism and their variations, see Adams, 1998).

Progressivism looks upon history (both cultural and biological) as following a natural law of continuous improvement in conditions and institutions. According to anthropologist William Adams, "No people, so far as I know, has ever propounded a progressivist doctrine without placing itself at the top of the ladder of progress. No wonder, then, that the doctrine flourishes in self-confident times: in pre-Socratic Greece, in the Enlightenment, in the Victorian age, and again in the America of the 1950s and 1960s." (Adams, p. 12)

Primitivism predates the ancient Greeks. The basis of the Garden of Eden myth and its Babylonian and Sumerian precursors, primitivism stems from "the nostalgia of civilized man for a return to a primitive or pre-civilized condition," the feeling that every step towards increasing cultural complexity brings with it "doubts about the whole enterprise of civilization." (Bell, 1972, p. 1). Primitivism, as Adams notes, "is more a commitment of the heart rather than of the head." (Adams, p. 78) It is a pessimistic doctrine, and frequently a form of self- (or societal-) castigation. "For the primitivist, the Golden Age was far in the past, and often at the very beginning of the world. Everything since has been a tale of increasing corruption of the originally pure state of nature, including Natural Man. In its most extreme form, Primitivism may involve a belief in the general superiority of animals over human beings—an ideology that can be recognized in some ancient Greek philosophy and again in the animal-rights movement of today." (Adams, p. 75)

While progressivists are cheerleaders for the inevitability of change, primitivists either deny that inevitability or are driven to despair by it. Progressivism is associated with cultural self-confidence; primitivism finds its home in times of cultural despair. Science takes a progressivist approach; primitivism is more welcome in the halls of religion and the arts. Media popularizations, such as the depiction of the Lakota Indians in Kevin Kostner's *Dances With Wolves*, owe more to the leitmotifs of primitivist literature than to historical research.

As anthropology developed in the 19th century it melded biological evolution with progressivism. It then became intertwined with the eugenics movement, nationalist politics, and the imperial "Great Game." Man rose "up from the ape," some races rose from savagery to barbarism to civilization, and some nations rose to become world empires. Why had some but not others ascended the ladder of progress? Since those doing the classifying were the ones asking the question, the answer was obvious and self-satisfying. Not everyone agreed, however, and the opening salvoes in the "anthropology wars" were fired in Britain, Germany, and the United States.

CHARLES DARWIN'S SMARTER, YOUNGER COUSIN

One of the first to seize on the implications of human variation was Darwin's younger cousin, Sir Francis Galton (1822-1911). Dubbed "a Victorian Genius," Galton authored over 300 publications, founded the scientific study of human differences, pioneered the use of fingerprints as a means of identification, and originated the twin method of genetic analysis and the use of correlational statistics. (Forrest, 1974)

Galton used Shakespeare's "alliterative antithesis" in the titles of his book *English Men of Science: Their Nature and Nurture* (1874) and his article in *Fraser's Magazine,* "The History of Twins, as a Criterion of the Relative Powers of Nature and Nurture" (1875). He wrote that "the phrase nature and nurture" provides "a convenient jingle of words, for it separates under two distinct heads the innumerable elements of which personality is composed. Nature is all that a man brings with him into the world; nurture is every influence from without that affects him after birth." (Galton, 1872, p. 12)

Galton believed his research had strong implications for education, criminology, economics, medicine and many other aspects of life. He was particularly concerned that the more moral and intelligent people were having fewer children than those less gifted. (Interestingly, his own long marriage produced no offspring). He therefore coined the word and inaugurated the science of eugenics, derived from the Greek "eugenes," for "good in stock, hereditarily endowed with noble qualities." (Forrest, p. 162 n.) Galton described two types of policies. Positive eugenic programs would give financial support to those deemed to be more intelligent to encourage them to have children. Negative eugenics aimed to reduce the fertility of those with severe intellectual, health, or character problems; to put it bluntly, to sterilize them. In 1908 (three years before his death), while reflecting on his life and work, Galton wrote: "Man is gifted with pity and other kindly feelings; he has also the power of preventing many kinds of suffering. I conceive it to fall well within his province to replace Natural Selection by other processes that are more merciful and not less effective. This is precisely the aim of Eugenics." (Galton, 1908, p. 323)

Galton also believed that there were race differences in intelligence. He estimated them by counting the number of eminent individuals they produced. His method was based on the statistical properties of the bell curve. A race that produced a high proportion of highly gifted individuals had a high level of intelligence. Using this method Galton constructed a 16-category scale of intellectual ability, ranging from mentally retarded to genius. He estimated that sub-Saharan Africans scored two categories below the English, and Australian aborigines one category below Africans. Each category corresponds to approximately 10 IQ points on a standard intelligence test. So, if the English were allocated an average IQ of 100, sub-Saharan Africans would have a mean IQ of about 80 and Australian aborigines of about 70. (See Figure 6.2)

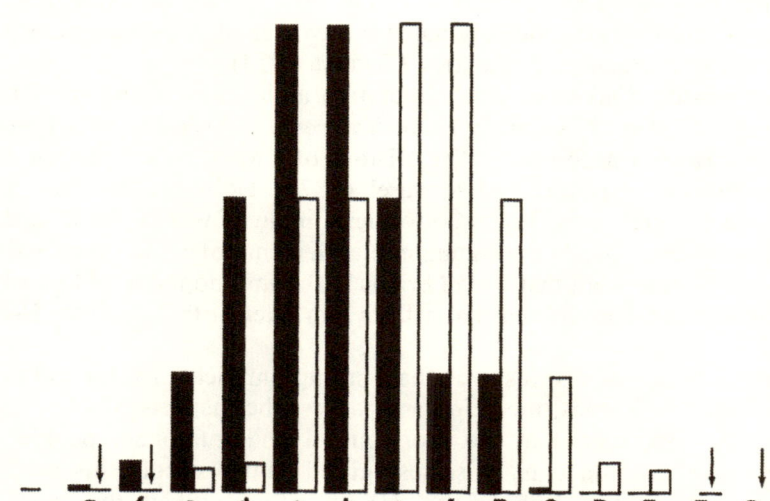

Figure 6.2: Galton's estimate of the range of mental ability in different races based on his tabulation of their number of eminent men. Right: The same data for Englishmen and Africans fitted to bell curves, showing both overlap and a difference in average intelligence.

Some of the problems with Galton's method are obvious: How accurate were his counts of eminent men? Does eminence mean the same thing in different cultures? Do all cultures even promote eminence as opposed to equality? In fairness to Galton, his method reveals average differences and considerable overlap between races. In this respect, it was more modern than the typological race theories of the early anthropologists. And for whatever it is worth, Galton's estimates of mean racial differences are not inconsistent with the results obtained with modern IQ tests. (Lynn, 1997) Nor was he out to prove that his fellow Victorian Englishmen were the pinnacle of evolution. Rather he concluded the Athenians of the 5th century BCE achieved the highest intelligence rating, two full grades above his contemporary countrymen:

> This estimate, which may seem prodigious to some, is confirmed by the quick intelligence and high culture of the Athenian commonality, before whom literary works were recited, and works of art exhibited, of a far more severe character than could possibly be appreciated by the average of our race, the calibre of whose intellect is easily gained by a glance at the contents of a railway bookstall. (Galton, 1892, p. 397)

One can only imagine what Galton might have thought of today's tabloid TV, dumbed-down textbooks, and sound bite political campaigns. A good guess comes from his less than flattering description of contemporary Americans:

> Enterprising, defiant and touchy; impatient of authority; furious politicians; very tolerant of fraud and violence; possessing much high and generous spirit, and some true religious feeling, but strongly addicted to cant. (Galton, 1892, p. 328)

EIN VOLK, EIN REICH, EINE PHILOSOPHIE

Darwinism in Britain, whether in the early days or today, has focused on individuals, with groups emerging from them. British evolutionism has always had the shopkeeper's sober obsession with keeping a good set of books. In Germany however, Darwinism took on a collectivist, romantic tone. There the great apostle of Darwin, Ernst Haeckel (1834-1913), imbued the theory of natural selection with the spirit of German Romanticism. The latter is hard to define, but easy to experience—just look at some paintings by Caspar David Friedrich while listening to a Wagner opera.

Haeckel and all he came to champion were opposed by his former professor, the distinguished biologist Rudolf Virchow (1821-1902). (For more on the conflict between Haeckel and Virchow, see Shipman, 1994; and Wolpoff and Caspari, 1997). The conflict between them was both personal and political. The two men were polar opposites in appearance, ancestry, and temperament. Haeckel was tall, blond, German in name and appearance, with a strong love of the outdoors, and a generalist looking for one grand theory to account for everything. Virchow, whose name and appearance betrayed a Slavic ancestry, was a detail man and a pedantic laboratory task master. Haeckel was charismatic and developed a huge, almost religious following; Virchow was respected, even feared, but rarely liked. Haeckel was a strong supporter of the German *Volk und Reich*; Virchow a radical advocate of social reform who fought at the barricades in the revolution of 1848. Virchow was a member of the German Progressive Party and opposed Bismarck's policies. The Iron Chancellor, having already dispatched or intimidated earlier opponents with sabre or pistol, challenged the professor to a duel. Virchow declined—unless they agreed to fight with scalpels. (Shipman, p. 89)

In 1860 Haeckel translated Darwin's *Origin* into German. His own master work, *Die Weltträtsel* (*The Riddle of the Universe*), used natural selection as the master key to explain all existence from inanimate objects, through a progression of animals, to human races. It was translated into 25 languages and sold over 100,000 copies in its first year, and eventually over half a million in Germany alone. Haeckel later termed the doctrine of *Die Weltträtsel*, Monism. It was pro-eugenic, Nordicist, nationalist, secularist, and hierarchical. In 1906 he founded the German Monist League to further the political application of its principles. According to historian Daniel Gasman, Haeckel's Monism was a direct precursor to National Socialist and Fascist ideology. (Gasman, 1971; 1998) In fairness it should be noted, however, that many Monists were liberals, drawn to the philosophy by its anticlerical aspects, and that the League "disbanded in 1933 rather than become 'coordinated' into the Nazi state." (Wolpoff and Caspari, p. 375, n. 73)

Between 1863 and Virchow's death in 1902, Haeckel and his former professor clashed at scientific conferences and in print. Haeckel's evolutionism was progressive, going from lower to higher forms. Without any physical evidence, Haeckel went out on a limb and predicted fossil hunters would soon discover a creature he dubbed *Pithecanthropus*, the apeman missing link. Inspired by Haeckel's prediction one of his disciples, Eugène Dubois, found the fossil he termed *Pithecanthropus erectus* (now classified as *Homo erectus*) in Java in 1891. For Virchow this was pointless speculation. He rejected the fossils, saying they were the result of pathological degeneration. As his repugnance at what he saw as the associations and implications of Monism grew, Virchow came to reject evolution altogether. Any change in individuals or species that we could observe rather than hypothesize, he argued, was evidence of degeneration, not progress. When Haeckel died in 1913 Monism had taken firm root in German soil. In the view of anthropologist Pat Shipman:

Between them, Virchow and Haeckel defeated empirical science in Germany altogether. By using science as the weapon of political reform, the one [Virchow] was led to deny the existence of evolution apparent to his eyes and the other [Haeckel] to mutate, expand, and wrench Darwin's poor theory out of all recognition. (Shipman, p. 103)

Although slightly exaggerated, Shipman's summary shows how dark a shadow Caliban has cast over anthropology—even to today. As we shall see, the Haeckel-Virchow dispute would not be the last time an in-group—out-group clash lurked beneath the surface of the anthropology wars.

GOOD-BYE RACE, HELLO CULTURE!

When Galton died in 1911, eugenics was widely accepted not only in Britain and Germany, but in the United States as well. Raymond Pearl, professor of biology at Johns Hopkins University (then a supporter of eugenics, but later an opponent), noted that by 1912, "eugenics was catching on to an extraordinary degree with radical and conservative alike." (Pearl, 1911, p. 335) Enthusiasts included literary giants H. G. Wells, George Bernard Shaw, and H. L. Mencken; crusaders for reproductive rights and sexual freedom Margaret Sanger and Havelock Ellis; scientists Harold Laski, J. B. S. Haldane, Alexander Graham Bell, and Luther Burbank; conservationist Gifford Pinchot; Winston Churchill (one of the English Vice Presidents of the First International Congress for Eugenics held in London in 1912) (Torrey, 1992, p. 47); socialist organizer Emma Goldman; Stanford University president David Starr Jordan and American Museum of Natural History president Henry Fairfield Osborn. (Kelves, 1985, pp. 21, 64) In 1918 Osborn joined biologist Charles Davenport and Madison Grant in founding the Galton Society for "the promotion of study of racial anthropology, and of the origin, migration, physical and mental characters, crossing and evolution of human races, living and extinct." (Quoted in Freeman, 1996, p. 48)

In 1927 Supreme Court Justice Oliver Wendell Holmes, arguably America's most brilliant jurist and hardly a conservative icon, supported state-mandated sterilization of the mentally retarded in the *Buck v. Bell* decision. Writing for an 8-to-1 majority that included noted civil libertarian Louis Brandeis, Holmes penned the immortal line, "three generations of imbeciles are enough." (Quoted by Degler, 1991, p. 47) *Buck v. Bell* has never been overturned by any subsequent Supreme Court decision. It was even cited by Justice Thurgood Marshall, as liberal a justice as ever to sit on the high bench, as "the initial decision," then reaffirmed by the famous *Roe v. Wade* abortion decision, that the Constitution provided no special protection for procreation. (Degler, pp. 47 - 48)

At the start of the 20th century, most American anthropologists came from wealthy "Brahmin" families and were educated at Harvard University. They were solidly in the eugenics camp, agreeing with Galton on both individual and race differences. And then, as one author put it, *Along Came Boas*. (Darnell, 1998; see also Cole, 1999, the first in a two-volume biography) While his name is hardly a household word, it is no exaggeration to say that Franz Boas (1858-1942) remade American anthropology in his own image. Through the works of his students Margaret Mead, Ruth Benedict, and Ashley Montagu, Boas would have more effect on American intellectual thought than Darwin. For generations, hardly anyone graduated from an American college or university without having read at least one of these authors' books. They all drew their inspiration from Boas's book *The Mind of Primitive Man*.

Franz Boas came from a German Jewish home steeped in the "sentiment of the barricades" of the 1848 revolutions that swept across Europe. He originally obtained his doctorate in physics but later turned to geography. Then, after field work with the Greenland Eskimos (now properly termed, the Inuit), he took up anthropology—Virchow's brand, not Haeckel's. While not religious, Boas was highly sensitive to anti-Semitism. His admiration for Virchow did not prevent him from fighting duels, including one that arose from an anti-Jewish slur. Repulsed by the rising tide of anti-Semitism in Bismarck's unified Germany, Boas left the fatherland he no longer felt to be his own and came to America.

Appointed chairman of the department at Columbia University in 1899, Boas transformed anthropology from the leisurely study of a few well-to-do WASPs into a highly credentialed discipline that pumped out Ph.D.s. By 1915 his students had a two-thirds controlling majority on the Executive Board of the American Anthropological Association. (Stocking, 1968; see also Stocking, 1987) In 1919 Boas could boast that "most of the anthropological work done at the present time in the United States" came from his former students at Columbia. (Stocking, 1968, p. 296) By 1926 they headed every major department of anthropology in America.

Before Boas, anthropology was the study of race. After Boas, anthropology was the study of culture, defined as "personality writ large," that is, "how a given temperamental approach to living could come so to dominate...that all who were born in it would become the willing or unwilling heirs to that view of the world." (Quoted in Stocking, 1989, p. 226, n. 6) In *Sex and Temperament in Three Primitive Societies*, Margaret Mead described how very different sex roles were among the Arapesh, the Mugdugumor, and the Tchambuli, three New Guinea peoples that lived within 100 miles of each other. She concluded that, "many, if not all, of the personality traits which we have called masculine or feminine are as lightly linked to sex as are clothing, the manners, and the form of head-dress that a society at a given point assigns to either sex." (Mead, 1950, p. 206) Ruth Benedict's *Patterns of Culture* contrasted what she dubbed the "Apollonian" (sober, egalitarian, and cooperative) Zuñi Indians of New Mexico with the "Dionysian" (excessively emotional, individualistic, and megalomaniacal) Kwakiutl of British Columbia and the "paranoid" (intensely jealous, suspicious, and resentful) Dobuans of New Guinea. These differences were cultural because "the biological bases of behavior in mankind are for the most part irrelevant." (Benedict, 1949, p. 206; for a critique, see Freeman, 1996; Stocking, 1968; and Torrey, 1992) As "Papa Franz" (as his students called him) had written in the foreword to Mead's *Coming of Age in Samoa*, "much of what we ascribe to human nature is no more than a reaction to the restraints put upon us by our civilisation." (Boas in Mead, 1949, x) In innumerable editions of *Man's Most Dangerous Myth: The Fallacy of Race*, Ashley Montagu waged a campaign to replace the term "race" with "ethnic group."

Historian Carl Degler (1991) emphasizes the essential role Boas played in decoupling the social sciences from biology: "Boas's influence upon American social scientists in matters of race can hardly be exaggerated." He engaged in a "life-long assault on the idea that race was a primary source of the differences to be found in the mental or social capabilities of human groups. He accomplished his mission largely through his ceaseless, almost relentless articulation of the concept of culture." (p. 61)

Like his mentor Virchow, Boas was skeptical of evolutionary explanations, genetic or cultural, even entertaining Lamarckism. What turned him into the godfather of cultural determinism in America, however, was the growing popular appeal and political power of the eugenics and anti-immigration movements. Standardized intelligence tests had recently been developed and administered to military recruits in World War I. Armed with results that showed large race differences in IQ, books like Madison Grant's *The Passing of the Great Race* (1921) argued that the survival of America depended on limiting immigration to those of Northwestern European descent. Southern and Eastern Europeans, and especially Jews, need not apply.

Franz Boas was a dark-haired Jewish immigrant from a leftist milieu, educated at German universities steeped in the ideals of the Enlightenment. Madison Grant, an archetypal Nordic, was a lawyer turned amateur biologist, and pillar of America's WASP establishment. Grant claimed that his fellow American Nordics were committing racial suicide, allowing themselves to be "elbowed out" of their own land by ruthless, self-interested Jewish immigrants. (pp. 16, 91) In the words of Yogi Berra, "it was déjà vu all over again." Haeckel's Monism had driven Virchow from skepticism into rejecting biological evolution. Nativist, pro-eugenic, elitist tracts such as Grant's drove Boas to reject the evolutionary perspective on culture and even linguistics, which he had earlier advocated (Sarich, 1994).

WHAT'S IN A NAME?

The early anthropologists planted their flag at the far right front summit of Caliban's Island. Human nature could be explained in terms of the universal principle of genetically based, progressive evolution. The Boasians snatched anthropology's flag and hauled it not only to the left (cultural) side, but also down into relativism and rearward toward primitivism. Their method was to cast doubt on general theories of human nature and instead emphasize diversity and the arbitrary nature of any standards of evaluation. Each culture simply had to be described, not explained. Any universal theory had to await a thorough cataloguing of cultural diversity. In fact, no general theory emerged from anthropology's Boasian half century. (Stocking, 1968, p. 210) Boas's theory of cultural determinism, or as Marvin Harris (1968, p. 250) terms it, Historical Particularism, would leave anthropology without any unifying principle. By rejecting the scientific validity of generalization and classification, Boasian anthropology was more an anti-theory than a theory of human nature. (White, 1966, p. 15)

While the Boasians considered the views of the eugenicists and cultural evolutionists to be value-laden (which is true), the works of Mead, Benedict, and the others were hardly value-neutral. Rather, what lurked below the surface were primitivist, moral indictments of contemporary Western society, especially its sexual mores.

The Boasians were outsiders. "Papa Franz" and many of his students were Jews, though "the preponderance of Jewish intellectuals in the early years of Boasian anthropology and the Jewish identities of anthropologists in subsequent generations have been downplayed in standard histories of the discipline." (Frank, 1997, p. 731; for more on the importance of Jewish identity in Boasian anthropology, see Hauschild, 1996, and MacDonald, 1998) Some, like Boas himself, were immigrants to boot. Montagu was born Israel Ehrenberg in the working class East End district of London. He was so leery of anti-Semitism ("If you're brought up as a Jew, you know that all non-Jews are anti-Semitic.... It's a good working hypothesis") that he re-invented himself as Montague Francis Ashley-Montagu from the well-to-do West End City of London financial district, complete with a posh public school accent. When he came to the United States, Montagu played the role of the British headmaster, lecturing American audiences before a receptive media on the foolishness of their prejudices. Later he dropped the hyphen and became simply Ashley Montagu. (Shipman, pp. 166, 159-160, 180-181)

Mead and Benedict could point to WASP pedigrees as pure as Madison Grant's, but Mead was bisexual and Benedict lesbian at a time when those orientations were far more stigmatized than they are today. (Torrey, pp. 60-83)

The Boasians shared an out-group sensibility, a commitment to a common viewpoint and a program to dominate the institutional structure of anthropology. (Stocking, 1968, pp. 279-280) Through it they successfully dethroned "the moral and political monopoly of an elite which had justified its rule with the claim that their superior virtue was the outcome of the evolutionary process." (Wolf, 1990, p. 168) The cultural determinism of the Boasians, emphasizing the variation within races, the overlap between them, and the plasticity of human behavior, served as a corrective to the genetic determinism of racial anthropology. By decoupling the science of man from the evolutionary perspective, however, they left anthropology sliding down the slippery slope of relativism, in danger of being marooned on the shore of primitivism or plunged into the abyss of deconstuctionism.

According to Degler "Boas, almost single-handedly, developed in America the concept of culture, which, like a powerful solvent, would in time expunge race from the literature of social science." (Degler, p. 71) In fact, Boas achieved his goal only with help, including a lot of it from a most unwelcome source—Hitler and the Holocaust.[i] Following W.W.II, "race" and "eugenics" became very dirty words. The University of London's Department of Eugenics changed its name to the Department of Genetics, the Eugenics Society became the Galton Institute, the Annals of Eugenics was renamed the Annals of Human Genetics, and Eugenics Quarterly became Social Biology. In 1949 UNESCO was tasked to adopt "a program of disseminating scientific facts designed to remove what is generally known as racial prejudice." The resulting 1950 UNESCO *Statement on Race*, written by Ashley Montagu, elevated the views expressed in *Man's Most Dangerous Myth* to the level of dogma.

As America moved into the Fabulous Fifties, the United States was engaged in a Cold War with the Soviet Union for "the hearts and minds" (and more importantly economic control) of the Third World. The American power elite, which was also becoming less and less a WASPs-only country club, realized the need to expunge Uncle Sam's racist record and rhetoric. A multi-racial United States was portrayed as having first saved the world from Hitler and now protecting it from Stalin. At the same time, government grants replaced wealthy benefactors as the principal source of anthropology's funding. (Coon, 1981, p. vii; Wolpoff and Caspari, p. 171) He who pays the piper calls the tune and the first law of politics is don't offend anyone or any group. Education, rather than social position, became the key to success. The burgeoning educational bureaucracy could best justify its existence by giving primacy to nurture and rejecting nature. Even the free market had a vested interest in nurturing nurturism. Advertising in Baby Boom America promoted the belief that by buying their kids "the breakfast of champions" or *The Book of Knowledge*, parents guaranteed themselves a seat at their scholar-athlete's college graduation.

The contrast between this paradigm and that of the previous century can be seen in two credos, the first from Francis Galton, the hereditarian and founder of eugenics:

> I have no patience with the hypothesis occasionally expressed and often implied, especially in tales written to teach children to be good, that babies are born pretty much alike, and that the sole agencies in creating differences…are steady application and moral effort. It is in the most unqualified manner that I object to pretensions of natural equality. (Galton, 1892, p. 14)

Contrast it with this from John B. Watson, the founder of behaviorism:

> Give me a dozen healthy infants, well-formed, and my own specified world to bring them up in, and I'll guarantee to take any one at random and train him to become any type of specialist I might select—doctor, lawyer, artist, merchant-chief, and, yes, even beggar-man or thief, regardless of his talents, penchants, tendencies, abilities, vocation, and race of his ancestors. (Watson, 1925, p. 270)

EVOLUTION STRIKES BACK

Cultural relativism placed anthropology in a theoretical vacuum, something scientific minds abhor. Newer, more sophisticated theories of the evolution of cultural complexity have seized that intellectual empty space. Massive data bases, such as George Murdock's Human Relations Area Files, a 450,000-page indexed catalogue of uniform descriptions of over 200 cultures, now allow them to be tested.

Figure 6.3 shows anthropologist Robert Carneiro's (1968) tabulation of the presence or absence of 50 cultural traits in 100 societies. Going from left to right we see that the Tasmanians and the Bambuti pygmies possess none of these 50 traits (which is not say they do not have a culture), while the Ancient Egyptians possessed all of them. Going from bottom to top on the list of traits we see that if a culture possesses any one of them, the odds are it possesses most if not all of those below. While not perfect, the figure shows a clear trend line of increasing cultural complexity. This is just what an evolutionary model would predict. Cultural Relativism and Historical Particularism, on the other hand, could have predicted only helter skelter.

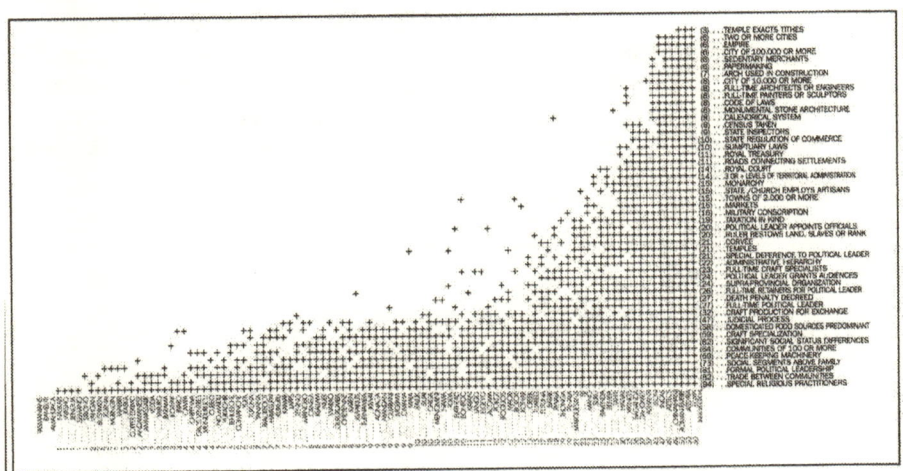

Figure 6.3: Carneiro's Scalogram of 50 cultural traits in 100 societies, showing a clear trend line of increasing complexity. This fits evolutionary models (genetic, environmental or both), but not cultural relativism which would predict only a random distribution of points.

The data are in, but the big questions are *why* are some societies more complex than others? And *what* causes cultural change? (For interesting discussions of the relation between race and cultural complexity, compare: Rushton (2000) [pro] against Diamond (1997) [con]; and on the relationship between race and IQ, contrast the interview of *Bell Curve* author Charles Murray (pro) in Chapter 10 with that of *Encyclopedia of Intelligence* Editor Robert Sternberg (against) in Chapter 11.)

Cultural evolutionary theory is intellectually alive and well and cultural relativism is off the theoretical radar screen. Some of today's anthropologists have climbed back up to the summit of Caliban's Island and planted their flag in much more solid empirical ground.

Twin and adoption studies, the Human Genome project, and the Decade of the Brain have established that both heredity and environment play a part in the development of just about every human behavior. Baby Boomers, now at mid-life, are anxious to stave off the diseases of old age. Through their elected representatives, government is now funding the search for gene therapies. Eugenic methods such as genetic screening have acquired a certain respectability provided that parents, not government, have control.

When modern multivariate statistical methods are applied to a series of measurements with large, representative samples of skulls from around the world (rather than just looking at the one composite measure of Cephalic Index) they produce pretty much the same picture as blood group, as well as the newer mitochondrial DNA and Y-chromosome techniques. Figure 6.4 shows two representations of human variation. The first based on comparing 120 separate genes in 38 populations; the second on Howell's multivariate statistical analysis of 57 cranial measurements in 28 groups. (Howells, 1993, p. 212)

GENETIC CLUSTER 42 GROUPS

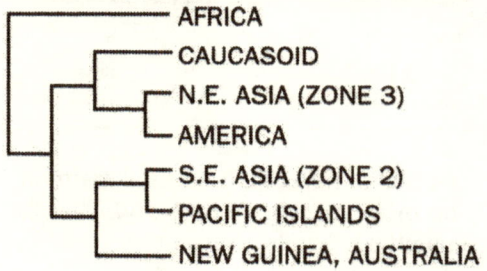

- AFRICA
- CAUCASOID
- N.E. ASIA (ZONE 3)
- AMERICA
- S.E. ASIA (ZONE 2)
- PACIFIC ISLANDS
- NEW GUINEA, AUSTRALIA

CRANIAL CLUSTER 28 GROUPS

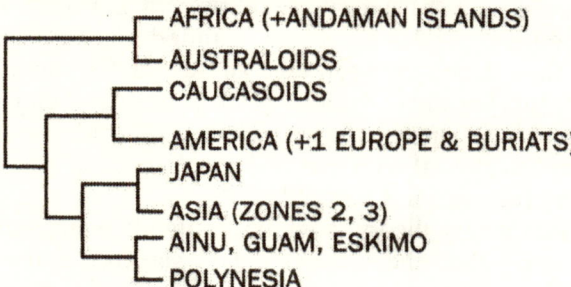

- AFRICA (+ANDAMAN ISLANDS)
- AUSTRALOIDS
- CAUCASOIDS
- AMERICA (+1 EUROPE & BURIATS)
- JAPAN
- ASIA (ZONES 2, 3)
- AINU, GUAM, ESKIMO
- POLYNESIA

Figure 6.4: Howell's comparison of genetic measures (top) and old fashioned measurements of skulls (bottom), showing that they paint very similar pictures of human origins and diversity.

Today, the early attempts of anthropology to link language and race seem ludicrous. There is no hereditary link between physical appearance or DNA on the one hand, and the language a person or group speaks, only a species-wide capacity to acquire any language during childhood. (Shermer, 2001). Conquest and immigration have caused whole groups to change their language. But as anyone who has tried to learn another language after childhood becomes painfully aware, it isn't easy. Language is cultural and arbitrary, but it is "sticky." Rob Foley's (1991, p. 114) measures of linguistic diversity paint a third picture that confirms those obtained from genetic diversity and multivariate analysis of crania. (See Figure 6.5).

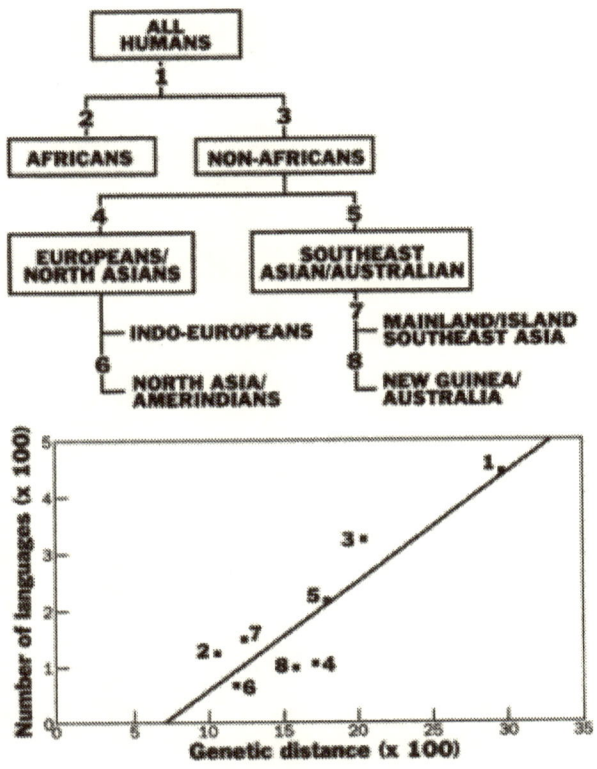

Figure 6.5: Foley's plot of genetic diversity in different populations (top) against the number of languages they speak (bottom). The nearly straight line shows that language also paints a similar picture of human origins and diversity.

All three methodologies support the "Out of Africa" model of recent, common, human origins. They lend no support for the early racial anthropologists' concept of static, Platonic essences, or types. Races, or populations[ii] as they are now euphemistically termed, are now understood as extended families, dynamic groups with differential gene frequencies that overlap and blend, but are still distinguishable. Human differences in gene frequencies, physical measurements, and to some degree historical languages trace the path of our common human origin. If they didn't it would argue for creation and against human evolution.

MAROONED ON CALIBAN'S ISLAND:
A DISCIPLINE IN SEARCH OF A DEFINITION

The most intellectually productive recent influence on anthropology came not from an authority on *anthropos*, but from an expert on ants—biologist Edward O. Wilson. What Boas had rent asunder, the final chapter of Wilson's (1975) *Sociobiology* re-joined.[iii] Human social behavior, including the phenomena of culture, could only be understood through the synthetic theory of evolution and concepts such as inclusive fitness and reciprocal altruism. The ensuing "Sociobiology Wars" are different from the "Anthropology Wars," (Segersträle, 2000; see also Part I, Chapters 1 through 5), but it is sufficient to note that Sociobiology set off a chain reaction among evolution-minded anthropologists.

One of the most significant of these was Napoleon Chagnon, whose research among the Yanomamö of the Amazon rain forest demonstrated that 40% of all adult males were complicit in the killing of another Yanomamö, 30% of all males had met violent deaths, and most importantly (and controversially), that killers left more offspring than other males. Chagnon's article hit Caliban's Island at ground zero—savages weren't so noble after all (with murder rates exceeding that of our own inner cities), and cultural success produced biological success. (Chagnon, 1998). Yanomamö homicide rates are not the exception. Most other tribal peoples are hardly peaceful and have equal or higher levels of killing (Keeley, 1996). Neither are traditional societies all harmonious, well-adapted (Edgerton, 1992), or environmentally-conscious (Krech, III, 1999) The majority of anthropologists, however, held strong in the faith of nurturism, relativism, and primitivism, and tried to have their discipline declared a sociobiology free zone.

The theoretical and empirical initiative has now shifted back toward the evolutionary perspective of nature (more accurately, gene-environment interaction), progressivism, and universalism. Most members of the American Anthropological Association, however, remain frozen in an ideological time warp. The most radical anti-evolutionists among them have sought refuge in the mire of deconstructionism, an exercise in literary solipsism that is more concerned with analyzing the "true meaning" of scientific "text" than the reality it attempts to describe and explain. With little agreement as to its subject matter or methods, anthropology today is a discipline in search of a definition, marooned on the back side of Caliban's Island, still haunted by the shadow of that "born devil, on whose nature nurture can never stick."

References

Adams, W. 1998. *The Philosophical Roots of Anthropology.* Stanford, CA: Center for the Study of Language and Information.

Baker, J. R. 1974. *Race.* New York: Oxford University Press.

Bell, M. 1972. *Primitivism.* London: Methuen.

Benedict, R. 1934. *Patterns of Culture.* New York: Mentor.

Boas, F. 1949. in Mead, M. 1949. *Coming of Age in Samoa.* New York: Mentor, 206, x.

Carneiro, R. 1968. "Ascertaining, Testing, and Interpreting Sequences of Cultural Development." *Southwest Journal of Anthropology.* Vol. 24, No. 4, 354.

Chagnon, N. 1988. "Life Histories, Blood Revenge, and Warfare in a Tribal Population." *Science.* 239: 985-992.

Cole, D. 1999. *Franz Boas: The Early Years: 1858-1906.* Seattle, WA: University of Washington Press.

Coon, C. S. 1939. *The Races of Europe.* New York: Macmillan, 107-108; Plate 5, Figure 1; and Plate 27, Figure 3.

Coon, C. S., Montagu, A., et al. 1950. "Race Concept and Human Races" in *Cold Spring Harbor Symposium,* Volume XV. Pp., 247-354.

Coon, C. S. and Hunt, E. E. Jr. (eds.) 1963. *Anthropology A to Z.* New York: Grosset & Dunlap, 133.

Coon, C. S. 1981. *Adventures and Discoveries: The Autobiography of Carleton S. Coon, Anthropologist and Explorer.* Englewood Cliffs, NJ: Prentice Hall, vii. Wolpoff, M. Op. cit., 171.

Darnell, R. 1998. *And Along Came Boas: Continuity and Revolution in Americanist Anthropology.* (Philadelphia PA; John Benjamin's Studies in the History of Linguistic Science Vol. 86.1998.

Darwin, C. 1874. *The Descent of Man and Selection in Relation to Sex.* London: John Murray.

Degler, C. 1991. *In Search of Human Nature.* New York: Oxford University Press.

Diamond, J. 1997. *Guns, Germs, and Steel.* W. W. Norton.

Edgerton, R. B. 1992. *Sick Societies: Challenging the Myth or Primitive Harmony.* New York: Free Press.

Foley, R. 1991. "The Silence Of the Past." *Nature.* Vol. 353. p.114.

Forrest, D. W. 1974. *Francis Galton: The Life and Work of a Victorian Genius.* New York: Taplinger.

Frank, G. 1997. "Jews, Multiculturalism, and Boasian Anthropology," *American Anthropologist*, Vol. 99, No. 4, 731.

Freeman, D. 1996. *Margaret Mead and the Heretic*. New York: Penguin, 48.

Galton, F. 1874. *English Men of Science: Their Nature and Nurture*. London: MacMillan.

Galton, F. 1875. "The History of Twins, as a Criterion of the Relative Powers of Nature and Nurture." *Fraser's Magazine*. Vol. 12, 566-576.

Galton, F. 1892. *Hereditary Genius* (2nd Edition). London: MacMillan.

Galton, F. 1908. *Memories of My Life*. London: Methuen.

Gasman, D. 1971. *The Scientific Origins of National Socialism*. London: MacDonald.

Gasman, D. 1998. *Haeckel's Monism and the Birth of Fascist Ideology*. New York: Peter Lang.

Gould, S. J. 1996. *The Mismeasure of Man*. Revised and Expanded Edition. New York: W. W. Norton & Company.

Grant, M. 1921. *The Passing of the Great Race; or the Racial Basis of European History*. New York: Scribner's.

Harris, M. 1968. *The Rise of Anthropological Theory*. New York: Crowell.

Hauschild, T. "Christians, Jews, and the Other in German Anthropology." *American Anthropologist*, Vol. 99, No. 4. 746-753; and 259-281.

Howells, W. 1993. *Getting Here: The Story Of Human Evolution*. Washington, D. C.: Compass Press.

Kevles, D. 1985. *In the Name of Eugenics*. New York: Knopf.

Keeley, L. H., 1996. *War Before Civilization: The Myth of the Peaceful Savage*. New York: Oxford University Press.

Krech III, S. 1999. *The Ecological Indian: Myth and History*. New York: W. W. Norton.

Lumsden, C., Wilson, E. O. 1981. *Genes, Mind, and Culture: The Coevolutionary Process*. Cambridge, MA: Harvard University Press.

Lynn. R. 1997. "Geographical Variation in Intelligence" in Nyborg, H. (ed.) *The Scientific Study of Human Nature: Tribute to Hans Eysenck at Eighty*. New York: Pergamon.

MacDonald, K. 1998. *Culture of Critique: An Evolutionary Analysis of Jewish Involvement in Twentieth Century Intellectual and Political Movements*. Westport, CT: Praeger.

Malia, M. 1999. "The Uses of Atrocity," in Courtois, S., et al. 1999. *The Black Book of Communism: Crimes, Terror, Repression.* Cambridge, MA: Harvard University Press, x.

Mayr, E. 1982. *The Growth of Biological Thought.* Cambridge, MA: Belknap, 567-568.

Mead, M. 1950. *Sex and Temperament in Three Primitive Societies.* New York: Mentor, 206.

Pearl, R. 1911. "Genetics and Eugenics," *Eugenics Review*, Vol. 3, 335.

Peterson, D. and Goodall, J. 2000. *Visions of Caliban: On Chimpanzees and People.* Athens, GA: University of Georgia Press.

Rushton, J. P. 2000. *Race, Evolution, and Behavior.* (3rd ed.). Port Huron, MI: Charles Darwin Research Institute.

Salter, F. 2001. Personal communication, 8 March.

Sarich, V. M. 1994. "Occam's Razor and Historical Linguistics." In Chen, M. Y. and Tzeng, O. J. L. In Honor of William S-Y. Wang (eds.) *Interdisciplinary Studies on Language and Language Change.* Pyramid Press: Taiwan, 410-430.

Schwartz, B. 1987. *The Battle for Human Nature.* New York: W. W. Norton. 215

Segerstråle, U. 2000. *Defenders of the Truth.* New York: Oxford University Press.

Shermer, M. 2001. "The Pinker Instinct: An Interview with Steven Pinker." *Skeptic*, Vol. 9, No. 1, pp. 88-96.

Shipman, P. 1994. *The Evolution of Racism: Human Differences and the Use and Abuse of Science.* New York: Simon and Schuster.

Stocking, G. W. 1968. *Race, Culture, and Evolution: Essays in the History of Anthropology.* New York: Free Press, 285.

Stocking, G. W. 1987. *Victorian Anthropology.* New York: Free Press.

Stocking, G. W. 1989. "The Ethnographic Sensibilities of the 1920s and the Dualism of the Anthropological Tradition," in G. W. Stocking (ed.) 1989 *Romantic Motives: Essays on Anthropological Sensibility*, 226 n.6. Madison, WI: University of Wisconsin Press.

Torrey, E. F. 1992. *Freudian Fraud: The Malign Effect of Freud's Theory on American Thought and Culture.* New York: HarperCollins.

White, L. 1966. "The Social Organization of Ethnological Theory." Rice University Studies: *Monographs in Cultural Anthropology.* Vol. 52. No. 4, 15.

Wolf, E. R. 1990. "The Anthropology of Liberal Reform" in Caton, H. (ed.). *The Samoa Reader: Anthropologists Take Stock.* Lanham, MD: University Press of America, 168.

Wolpoff, M. and Caspari, R. 1997. *Race and Human Evolution: A Fatal Attraction.* New York: Knopf.

Washburn, S. 1951. "The New Physical Anthropology." *Transactions of the New York Academy Sciences.* Series 2, Vol. 13. No. 7, 298-304;

Washburn, S. 1950. Cold *Spring Harbor Symposium on Quantitative Biology.* "Origin and Evolution of Man." Volume XV. 1950. Cold Spring Harbor.

Watson, J. B., 1925. *Behaviorism.* New York: Norton.

Wilson, E. O. 1975. *Sociobiology: The New Synthesis.* Cambridge, MA: Belknap.

Wright, L B. (ed.)"A True Reportory of the Wreck and Redemption of Sir Thomas Gates, Knight," in 1964. *A Voyage to Virginia in 1609.* Charlottesville, VA: University of Virginia Press.

NOTES

1. The record of Hitler's Germany, which is frequently invoked in an attempt to quash any discussion of race and genetics, is well enough known and acknowledged that it needs no documentation here. The record of the system that attempted, in Stalin's words, to be "the engineer of human souls," communism, is far less so and merits a brief citation. "Communism was in truth a 'tragedy of planetary dimensions'…with a grand total of victims variously estimated by contributors to the volume at between 85 million and 100 million…the most colossal case of political carnage in history." Malia, M. 1999. "The Uses of Atrocity," in Courtois, S., et al. 1999. *The Black Book of Communism: Crimes, Terror, Repression.* Cambridge, MA: Harvard University Press, x. It is also worth noting in this context that in the 1930's when Stalin was firmly in power, Boas referred to the Soviet Union as "the greatest experiment that has ever been made in Social Science with a high ideal in view…." Caffrey, M. M. 1989. *Ruth Benedict: Stranger in This Land.* Austin, TX: University of Texas Press, 289. Quoted by Torrey, Op. cit., 57.

2. The term "race" has been used so imprecisely and carries such emotionally-loaded additional meaning, a strong case can certainly be made for dropping it. "Population," is similar in having many additional meanings, but opposite in that it carries almost no weight at all. For the life of me, I wish I could come up with a better term. In 1974, British biologist John Baker wrestled with the same problem and suggested in a footnote (p. 5 n.) the Latin term *stirps* (plural, *stirpes*), and hence *stirpal*, rather than "racial" or "populational" differences, but it had no traction and never caught on. Perhaps the concept of fractals could be helpfully employed here.

3. Sociobiology did not spring fully-formed from the brow of E. O. Wilson, however. The related tradition of ethology was kept alive in Europe, mainly with regard to animal studies. It gained great prestige in 1973 when Karl von Frisch, Konrad Lorenz, and Niko Tinbergen were awarded the Nobel prize in Physiology or Medicine. Lorenz and his student Irenäus Eibl-Eibesfeldt applied the ethological method to humans. Wilson told Frank Salter, a colleague of Eibl-Eibesfeldt, at the 1989 meeting of the Human Behavior and Evolution Society (HBES), that by adding population genetics to ethology, sociobiology allowed its theories to be tested empirically. The two traditions have now essentially merged. Whether you describe yourself as a human sociobiologist, a Darwinian anthropologist, or a human ethologist depends largely on the job you're applying for. (Salter, F. 2001. Personal communication, 8 March).

CHAPTER 7

HOW HUMANS
EXPLAIN HUMAN ORIGINS

THE BONES, (STONES, AND DIRTY STORIES) OF CONTENTION

The need to explain our origins is one of the universals of being human. Creation myths are based on cultural beliefs that have, in one manner or other, been adopted as legitimate explanation by a particular society. To a large extent, creation myths glorify the specialness of humans. In the broadest view, such myths undertake to explain our differences from all other creatures—our humaness.
—Donald Johanson and Blake Edgar, *From Lucy to Language*, p. 18.

I realized that I was standing in front of a genre of literature, that I could approach the study of human evolution as a study of literature.... The question to ask, then, is not what do fossils tell us about human evolution but what is it about human evolution—and not only human evolution—that through fossils is getting said?
—Misia Landau, quoted in Roger Lewin, *The Bones of Contention*, pp. 32, 43.

Evolutionary scenarios must be evaluated much in the same way that jury members must judge a prosecutor's narrative. Ultimately they must make their judgment not on the basis of any single fact or observation, but on the totality of the available evidence. Rarely will any single item of evidence prove pivotal in determining whether a prosecutor's scenario or the defense's alternative is most likely to be correct. Many single details may actually fail to favor one scenario over another. The most probable account, instead, is the one which is the most internally consistent—the one in which all the facts mesh together most neatly with one another and with the motives in the case. Of paramount importance is the economy of explanation. There are always

129

alternative explanations of any single, isolated fact. The greater the number of special explanations required in a narrative, however, the less probable its accuracy. An effective scenario almost always has a compelling facility to explain a chain of facts with a minimum of such special explanations. Instead the pieces of the puzzle should fall into place.

—C. Owen Lovejoy, "Modeling Human Origins: Are We Sexy Because We're Smart, Or Smart Because We're Sexy?", p.2.

Looking back on his life as one of its most important and controversial figures, Carleton Coon (1981, p. vii) wrote that anthropology was divided into three parts: "bones (physical anthropology), stones (prehistoric archaeology), and dirty stories (ethnology and ethnography)." Science writer Roger Lewin explained that the title of his book on the search for human origins, *The Bones of Contention*, was drawn from the title of the 1958 Huxley Memorial Lecture in which Sir Wilfred Le Gros Clark (another founding father of the discipline) declared, "Every discovery of a fossil relic which appears to throw light on connecting links in man's ancestry has, and always will, arouse controversy" (in Lewin, 1987, p.19).

To mix the metaphors of Coon and Le Gros Clark, the contentious history of how humans explain human origins is filled with not only bones, but also stones, and more than its share of dirty stories. Most dirty stories contain four-letter words. In the case of human origins, the four-letter word is R-A-C-E. Anthropologist Milford Wolpoff titled his recent popular book, *Race and Human Evolution: A Fatal Attraction*. But most fatal attractions, not only the one dramatized in the hit film, are to some extent written in our genes.

As Lewin states (p.28), the questions to be answered by origin stories—both scientific and non-scientific—are: "Who?, Where?, When?, and Why? Who was our ancestor? Where did it first arise? When did we break away? And, Why did it happen?" But as Landau's quote points out, an equally important set of questions seems to be "Who came up with that version of human origins? Where and when did they come up with it?" and most importantly, "Why did they come up with it?" While engaging in the sociology of knowledge can lead one down an infinite regress, explanations of human origins and human differences have been subject to more than their share of outside influences. And that piece of history goes back a long, long way. If Donald Johanson is correct, the need to explain our origins is one of the universals of being human, and creation myths that glorify the specialness of humans serve not only to explain our differences from all other creatures but to explain our differences from other humans, and that in itself may tell us something about human nature and evolution.

ORIGIN STORIES AND ETHNOCENTRISM

> In the beginning there was Jo-Uk, the Great Creator, and he made the Sacred White Cow. The White Cow gave birth to a man child whom she called Kola, whose grandson was Ukwa. Ukwa took two wives, dark virgins who also rose out of the holy water. One of Ukwa's son's Nyakang, a tall blue-black warrior, went south to the marshes of the Upper Nile; there he founded the Shilluk nation and became its first *ret*, or ruler, and a demigod. All this happened when the world was new, about four hundred years ago (see Freund, p. 5-6).

Examples of other non-Western origin stories can be found in works such as Freund's *Myths of Creation*, von Franz's *Creation Myths*, and other compendia. Note that these are not accounts of anything so academic as the evolution of humankind—they are accounts of the origin of the group—the tribe, the people, the nation, the race. They are stories that initiate members into the group by explaining what makes it so important and so different from other groups.

William Graham Sumner was perhaps the first to observe, in 1911, that "nine-tenths of all the names given by savage tribes to themselves mean 'men,' 'the only men,' or 'men of men'; that is, 'We are men, the rest are something else'." Further, he observed in 1906, "Each group nourishes its own pride and vanity, boasts itself superior, exalts its own divinities, and looks with contempt on outsiders. Each group thinks its own folkways the only right ones, and if it observes that other groups have other folkways, these excite its scorn" (in van der Dennen, p. 5). Modern anthropological research has confirmed that among relatively isolated peoples, whether horticulturists, pastoralists, or hunter-gatherers who have not been incorporated in a meaningful way into states, the word for one's group simply means people or humanity, in contrast to aliens, others, and "non-us." This terminology is a reflection on an underlying enthocentrism that varies from the simple expression of cultural pride at one end to a strong attempt to deny humanity to those who belong to different cultures at the other. Each group sees itself as special, unique, favored of God or the gods, and in some sense superior.

Recent evidence for the evolutionary basis of enthocentrism and its potential for evolution into nationalism and genocidal racism can be found in *The Sociobiology of Ethnocentrism* (Reynolds, Falger, and Vine [eds.],

1987), in van der Dennen's *The Origin of War*[1995], and in *On Human Nature* (1978) by the founder of sociobiology, E. O. Wilson. Hirschfeld's *Race in the Making: Cognition, Culture, and the Child's Construction of Human Kinds* (1996) presents evidence that the tendency to divide the world into basic "kinds" of us-versus-them (what Sumner termed in-group amity versus out-group enmity), is a basic process in human cognitive development.

As groups come into contact, origin stories can serve either to ease their integration into a single, larger group, or to maintain their separation. The dominant group clearly has an interest in convincing themselves that they deserve to be there and even more so in trying to convince subordinate groups that they should accept their position. Looking just at American history, one sees that interpretations of the Bible have been invoked to lend divine sanction to slavery, segregation, and the conquest of the continent. But as Michael Shermer (1997) points out in "God and the Ghost Dance," given just a little bit of rewriting, the same stories can be used not as opiates to numb the oppressed into accepting their oppression, but rather as stimulants to goad them into revolution.

REVERSAL OF MYTHICAL FORTUNE

The origin story of the Nation of Islam (NOI) provides a good example of such reworking. As summarized in *The Autobiography of Malcolm X* (Haley, 1966, pp. 164-7), 6,600 years ago all men were Black Africans. An evil scientist, Yacub, was exiled from Mecca, whereupon he engaged in a series of selective breeding experiments producing successively the brown, red, yellow, and white races, until he finally achieved his goal of a race of blond-haired, blue-eyed, white devils who dwelt in caves in Europe until Moses civilized them. Thereupon they ruled the world for another 6,000 years until Allah's messenger called the world to repentance to mark the beginning of the end of their evil reign. Given the use, not only of the Bible but of racial science to justify White supremacy, the NOI origin story is not only understandable, but almost predictable in the manner in which it reworks similar themes, but this time to give Blacks the position of priority, dignity, and supremacy.

Such reversal of mythical fortune did not, however, begin or end with the Nation of Islam. Consider the books of *Genesis* and *Exodus*, which rework earlier Sumerian and other Near Eastern mythological themes to make the Jewish people the apple of god's eye. While the Bible stories about Joseph going to Egypt and rising to prominence, the bondage under Pharaoh, and the deliverance by Moses, were engraved in the minds of previous generations and immortalized on celluloid for more recent ones by Cecil B. DeMille and company, the Egyptian archives and archaeological record provide no evidence whatsoever that the Hebrews were ever in Egypt (Stiebing, 1989; Redford, 1992). Akenson (1992) shows how this same story—a people make a covenant with God, who in return grants them a special piece of land which they, with His divine help, turn from a wilderness into a paradise—has been reworked to provide support for the cause of not only modern Israel (which also has a more recent origin novel and film in Exodus), but also of the South African Boers and of the Ulster Protestants of Northern Ireland.

For the ancient Greeks, Homer's *Iliad* provided an account of their own rise to prominence. Alas, when the upstart Romans achieved hegemony, they were without such an effective origin story. No problem. Virgil's *Aeneid* took a relatively minor character from the *Iliad*—Aeneas—and made him the founder of Rome (the new Troy). Now Roman world-unification and domination under Augustus would appear not as the spoils of bloody wars, but rather as a divinely ordained necessity. Odysseus (Ulysses) went from being the cunning hero who saves the day when all appears lost in Homer, to a virtual war criminal in the *Aeneid*. Since Virgil was financially supported by Maecenas, a member of Augustus' court, the *Aeneid* may be the first example of a state-sponsored origin myth!

Even that most self-consciously upstart of all nations, the United States of America, has its own justifying origin myth—*The Winning of the West*, in four volumes, written by no less than Theodore Roosevelt, later to be one of its most popular presidents. In popular culture, the story of the hero and, not the New Troy but the New Jerusalem, has been presented in innumerable movies and personified by that American icon, the Duke—John Wayne. Even subgroups within modern America have their own pop-culture origin stories, which tell of the rise of their hero from humble beginnings to a position of power in the brave, new world. For Italian-Americans, it's *The Godfather*, and its hero is Don Vito Corleone. Puzo's novel drew heavily from sociologist Francis Ianni's research on the real-life "Lupollo" Mafia family (see Ianni, 1972), the latter organization having begun in Sicily during the late Middle Ages as "a secret organization dedicated to overthrowing the rule of various foreign conquerors...Saracens, Normans, and Spaniards." In other words, the Mafia was formed as a national liberation front—at least that's their origin myth.

JUST HOW SCIENTIFIC ARE SCIENTIFIC ORIGIN ACCOUNTS?

Compared to such self-serving ethnic origin myths just how scientific are the "scientific" accounts of human origins? Certainly, many early evolutionary accounts tell a story where "Darwinian man is lord of earth, not because of any God-given stewardship or Romantic affinity to the World Spirit, but for the good and legitimate reason that the British were rulers of Africa and India" (Cartmill, p.68). Given this background and its persistence, it is predictable that some Native American activists might turn a skeptical eye toward anthropological accounts of the peopling of the Americas and suspect them of being tainted by an underlying agenda. Such biases and agendas are not unknown in the history of anthropology, as Derek Freeman documents in his account of "The Margaret Mead Controversy" in *Skeptic*. However predictable, it is still unfortunate that anyone would react, not by looking upon anthropological accounts with heightened scrutiny, but by instead embracing the home-grown creationist accounts, which Ken Feder in his article in this issue, "Native American Creationism," finds as devoid of any scientific basis as their respective Ancient Greek, Roman, Biblical, or Nation of Islam counterparts.

Anthropologists themselves are not unaware of the past weaknesses of their field. J. S. Jones wrote in the leading British science journal *Nature*, "Paleoanthropologists seem to make up for lack of fossils with an excess of fury, and this must now be the only science in which it is still possible to become famous just by having an opinion." Andrew Hill, also in *Nature*, noted the manner in which "every new fossil hominid specimen is the most important ever found and solves all known phylogenetic problems; every new hominid specimen is completely different from all previous ones, no matter how similar; every new hominid specimen is a new species, probably a new genus, and therefore deserves a new name." David Pilbeam concluded "there is an inverse relationship between familiarity [with the fossil record] and expressed certainty of opinion" (all quoted in Howells, 1993, p. 78).

Misia Landau (1984) applied the methodology developed in Vladmir Propp's *Morphology of the Folktale* (1925) to five classic accounts written in the early 20th century in which the authors (note that two were knighted) presented their first clear and complete view of human evolution: Sir Arthur Keith's "Man's Posture: Its Evolution and Disorders" (1923), Sir Grafton Elliot Smith's *The Evolution of Man* (1924), Frederick Wood Jones's *Arboreal Man* (1916), Henry Fairfield Osborn's "The Plateau Habitat of the Pro-Dawn Man" (1928), and William King Gregory's "Studies on the Evolution of the Primates" (1916).

Propp's method consists of analyzing different folk tales in terms of a constant set of functions (the actions that take place), rather than in terms of their varied dramatis personae. Applying this method to Darwin's account in *The Descent of Man* and the five subsequent accounts of human evolution listed above, Landau found that Piltdown man (accepted as legitimate at the time) played "several roles in human evolution, from missing link to the first Englishman" (much as Odysseus went from hero to bum in moving from Homer's *Iliad* to Virgil's *Aeneid*). However, she also found that the evolutionary accounts did share four basic events—(1) terrestriality (the shift from the trees to the ground), (2) bipedalism (the development of upright posture), (3) encephalization (the expansion of the brain and the development of intelligence and language), and (4) culture (the development of society, technology, and morals). The accounts differed considerably, however, in terms of the order in which they claimed these events took place.

Continuing the analysis, Landau argued that, like many myths, the accounts of human evolution begin with a state of equilibrium, where the hero leads a peaceful life, though he is somehow different (perhaps smaller, weaker, or from more humble origins) than the other creatures. By choice or compulsion, the hero is forced to leave this idyllic home (either by developing an enlarged brain in Smith's scenario, or assuming an upright posture according to Keith and Wood Jones) and embarks on a journey, where he encounters and passes a series of tests which reveal his true nature. In his adventures, the hero is assisted by some special gift (tools, reason, or a moral sense), is tested more thoroughly, only to emerge triumphant in the end.

THE STORY THAT TELLS ITSELF

So are the accounts of human evolution just so many "just so" stories? In the chapter that follows, "In Love With Lucy—and All Her Relatives," Donald Johanson, perhaps the world's best known living paleoanthropologist, decries the way the scientific account of human origins is still all too often portrayed in such scenarios of an inexorable "march through time in which a Miocene ape simply stepped into the 'Magic Tunnel of Evolution' and emerged as an upright, White, European male." He notes with pride, however, that paleoanthropology has succeeded in determining whether bipedalism preceded encephalization (Darwin ends up looking pretty good, as always), not by grand theorizing as in the early accounts analyzed by Landau, but by finding the hard evidence—enough actual fossils, to settle the issue. Johanson goes on to make the case that paleoanthropology needs to restrain itself from flights of grand theoretical fancy and instead stick to the hard task of finding still more fossils and then performing the necessary detailed analyses.

The latest high-tech analysis of Neandertal mitochondrial DNA (Krings, M. et al., 1997) may even settle the long-standing argument between Out-of-Africa, no Neandertals in our-gene pool advocates like Stringer (Stringer and McKie, 1996, Stinger and Gamble, 1993) and multiregionalist, pro-Neandertalers like Wolpoff (Wolpoff and Caspari, 1997) and Trinkaus (Trinkaus and Shipman, 1993). New absolute dating techniques such as thermoluminescensce and electron spin resonance can now be used to supplement and check the dates derived from isotope dating methods (such as radiocabon and potassium/argon) or relative dating methods (such as stratigraphic analysis). When the newer methods produce a different dating, the scientific accounts of human origins are changed accordingly (see Stringer and Gamble, pp. 58-59). This never happens with creationist stories, "scientific," or otherwise. Instead, their proponents go to great lengths to explain away any dates that conflict with their 'received view'. Even literary deconstructionists would find it hard to explain why the physicists who have developed these new dating methods end up supporting the paleoanthropological accounts unless they concede that both scientific "stories" reflect an underlying reality. As the epigram from Lovejoy points out, science is capable of resolving even historical issues to a reasonable degree of certainty.

Does the fact that Landau's analysis of evolutionary accounts shows them to follow the same pattern as many human fictional narratives forever impeach their scientific credibility?

How else could one tell such a story? Remember, evolution has shaped not only our bodies, but also our minds, including our artistic (Dissanayake, 1992) and literary productions (Carroll, 1995), religions and ideologies

(Boyd and Richerson, 1985; Hartung, 1995; MacDonald, 1994). Given that the fossil record has now proven that bipedalism preceded encephalization, the question of human evolution then becomes, "What drove that increasing encephalization—the three-fold increase in the size of the hominid brain?" The classic answer has been, physically mastering the environment—making tools to kill game and provide shelter. But recently, consideration has been given to the role of sociality, the possibility that we drove our own evolution (Byrne, 1996). Each increase in social organization provided an evolutionary payoff for increasing brain size, which in turn, led to increased social organization. And what do we do in social settings? Tell stories! Two-thirds of our conversation, whether between males, females, or males and females, business occasion or pleasure, is devoted to social topics; no other topic accounts for more than 10% (Dunbar, p. 123). So telling stories may not only be the product of human evolution, telling stories, especially dirty stories (the ones about love, hate, and all that), could also be part of the cause. The story of human evolution may well turn out to be a story that is telling itself.

Figure 7.0: There goes the neighborhood! Every time Cro-magnons move in it's graffiti, graffiti, graffiti…"

REFERENCES

Akenson, D., 1992. *God's People: Covenant and Land in South Africa, Israel, and Ulster. Ithaca*, NY: Cornell.

Boyd, R., and Richerson, P. *Culture and the Evolutionary Process.* Chicago: University of Chicago Press.

Byrne, R. 1996. "Machiavellian Intelligence," *Evolutionary Anthropology*. Vol. 5. No. 5, pp. 172-180.

Cartmill, M. 1983. "Four Legs Good, Two Legs Bad" *Natural History*. November: pp. 65-78.

Carroll, J. 1995. *Evolution and Literary Theory*. Columbia, MO: University of Missouri Press.

Coon, C. S., 1981. *Adventures and Discoveries: The Autobiography of Carleton S. Coon*. Englewood Cliffs, NJ: Prentice Hall.

Dissanayake, E. 1992. *Homo Aestheticus: Where Art Comes From*. New York: The Free Press.

Dunbar, R. 1996. *Grooming, Gossip, and the Evolution of Language*. Cambridge: Harvard University Press.

Freeman, D. "Paradigms in Collision: Margaret Mead's Mistake and What It Has Done to Anthropology" *Skeptic* Vol. 5, No. 3, p. 68.

Freund, P. 1965. *Myths of Creation*. New York: Washington Square Press.

Haley, A. 1966. *The Autobiography of Malcolm X*. New York: Grove Press.

Hartung, J., 1995. "Love Thy Neighbor: The Evolution of In-Group Morality." *Skeptic* Vol. 3, No. 4, pp. 86-99.

Hirschfeld, L. 1996. *Race in the Making: Cognition, Culture, and the Child's Construction of Human Kinds*. Cambridge: MIT Press.

Howells, W. 1993. *Getting Here: The Story of Human Evolution*. Washington, DC: Compass Press.

Ianni, F. 1972. *A Family Business: Kinship and Social Control in Organized Crime*. NY: Sage.

Johanson, D., and Edgar, B. 1997. *From Lucy to Language*. New York: Simon and Schuster.

Krings, M., Stone, K A. Schmitz, R. W., Krainitzki, H., Stoneking, M., & Paabo, S. 1997. "Neandertal DNA Sequences and the Origin of Modern Humans." *Cell*, 90: 19-30.

Landau, M. 1984. "Human Evolution as Narrative." *American Scientist*. Vol. 72. pp. 262-268.

Lewin, R. 1987. *The Bones of Contention*. New York: Simon and Schuster.

Lovejoy, O. 1993. "Modeling Human Origins: Are We Sexy Because We're Smart, or Smart Because We're Sexy?" in Rasmussen, T. D., *The Origin and Evolution of Humans and Humaness*. Sudbury, MA: Jones and Bartlett.

MacDonald, K., 1994. A People That Shall Dwell Alone: Judaism as a Group Evolutionary Strategy. Westport, CT: Praeger.

Propp, V. 1925. *Morphology of the Folktale*. Austin, TX: University of Texas Press.

Redford, D., 1992. *Egypt, Canaan, and Israel in Ancient Times*. Princeton, NJ: Princeton University Press.

Shermer, M. (1997) "God and the Ghost Dance," *Skeptic* Vol. 3, No. 4. pp. 65-73.

Stiebing, W. 1989. *Out of the Desert? Archaeology and the Exodus/Conquest Narrative*. Amherst, NY: Prometheus.

Stringer, C., and Gamble, C. 1993. *In Search of the Neanderthals: Solving the Puzzle of Human Origins*. London: Thames and Hudson.

Stringer C., and McKie, V. 1996. *African Exodus: The Origins of Modern Humanity*. New York: Holt.

Trinkhaus, E., and Shipman, P. 1993. *The Neandertals: Changing the Image of Mankind*. New York: Knopf.

van der Dennen, J. 1987. "Ethnocentrism and In-group/Out-group Differentiation: A Review and Interpretation of the Literature" in Reynolds, V., Falger, V., and Vine, I., *The Sociobiology of Ethnocentrism*. Athens, GA: University of Georgia Press.

van der Dennen, J. 1995. *The Origin of War*. Groningen, The Netherlands: Origin Press,

von Franz, M-L., 1995. *Creation Myths*.(Revised Edition). Boston: Shambala.

Wilson, E. O. 1978. *On Human Nature*. Cambridge, MA: Harvard University Press.

Wolpoff, M, and Caspari, R. 1997. *Race and Human Evolution: A Fatal Attraction*. New York: Simon and Schuster.

CHAPTER 8

IN LOVE WITH LUCY
AND ALL HER RELATIVES

AN INTERVIEW WITH
PALEOANTHROPOLOGIST DONALD JOHANSON

Donald Johanson is perhaps the world's best known living paleoanthropologist. He is the founder and president of the Institute of Human Origin, (which recently relocated from Berkeley, CA, to Arizona State University in Tempe, AZ, where he also teaches at both the graduate and undergraduate level), discoverer of the Lucy skeleton that revolutionized our knowledge of human evolution, host of the acclaimed three-part PBS-Nova documentary *In Search of Human Origins*, and the author of numerous scientific papers and five previous books: *Ancestors: In Search of Human Origins* (with Lenora Johanson and Blake Edgar); *Journey from the Dawn: Life with the World's First Family* (with Kevin O'Farrell); *Lucy's Child: The Discovery of a Human Ancesto*r (with James Shreeve); *Lucy: The Beginnings of Humankind* (with Maitland Edey); and *Blueprints: Solving the Mystery of Evolution* (with Maitland Edey).

Johanson's most recent book, *From Lucy to Language* (with Blake Edgar, and special photography by David Brill, Simon and Schuster, 1997), grew from his commitment, in this time of neo-Fundamentalism, to provide the general public with access to the evidence the experts use in trying to reconstruct the story of human evolution. It provides the reader with a veritable home museum of human evolution. In addition to the informative and stimulating text, it includes beautiful life-sized photographs of every major fossil hominid skull that seem to almost jump out from the pages. One can only pray for it appear on CD-ROM (perhaps supplemented by numerical appendices which could be the source for countless future research articles, theses, and dissertations). With the publication of *From Lucy to Language*, creationists who try to foster doubt as to the fact of human evolution are finding themselves in the position of the defense team in the OJ civil trial after the photos of the Bruno Magli shoes surfaced.

In this interview, Johanson tells *Skeptic* why he thought it was important to produce so special a book as *From Lucy to Language*, responds to the religious, scientific, and literary criticisms of the search for human origins, sets forth the scientific and philosophical foundations of paleoanthropology, provides a great quick start guide to the latest findings in the search for our ancestors, tells us what he has learned in over two decades of field research in the Great Rift Valley of East Africa, and also shares his firsthand insights into the personalities who have shaped the ever exciting and ever controversial study of human evolution.

Donald Johanson is clearly in love with the search for our origins, with Lucy and all of her relatives. Pick up a copy of *From Lucy to Language* and you will be too.

Figure 8.0: Donald Johanson

Miele: You've taken time out from digging in East Africa for Lucy's relatives, writing up your research for the professional journals, and running the Institute of Human Origins to bring us your latest book, *From Lucy to Language* (Johanson and Blake Edgar, with photography by David Brill, Simon and Schuster, 1997). In addition to the very readable and informative text, it includes remarkable photographs of what seems like every major fossil hominid skull. Why the commitment to popularization, in the best meaning of that word?

Johanson: We decided to produce this book mainly because there was nothing else available for undergraduates and the lay public that clearly presents the evidence paleoanthropologists use in trying to reconstruct the human evolutionary career. They often hear the criticism that we base our theories and conclusions about human evolution on fragmentary pieces of jaw or just a single tooth. *From Lucy to Language* contains life-sized, not postage stamp-size, photographs of what we consider the most critical and pivotal fossils that have shaped our understanding of the human evolutionary career. For the first time, the reader can look at the specimens in their absolute natural size and get an impression of the actual anatomy of these creatures. You can see how different a *Homo erectus* skull is from a Neandertal, or how a Neandertal skull differs from a modern human skull. We were especially fortunate in having the help of David Brill, who is the world's premier photographer when it comes to rendering these fossils in virtually three dimensions. Students have come to me recently and said, "I could almost reach out and touch a browridge or palpate the bone surface where a muscle is inserted." For those students and general readers who don't have access to the original fossils or even casts, our book provides a good substitute.

Miele: Let's examine those criticisms. On the one hand you have creationists who claim that, "If you study the history of evolutionary anthropology you'll see that at one time they claimed Piltdown man was important, but now concede that it's a fraud. And then they claimed that *Ramapithecus* was the earliest hominid, but now concede that it's really a fossil orangutan. All these bone peddlers do is change their line-up card as their supposed 'missing links' get successively ejected from the evolutionary game."

On the other hand, you have molecular evolutionists who say, "We already know great deal from the comparative data, both anatomical and molecular. How would our knowledge of our evolution suffer if there was no hominid fossil record at all?" And most recently, we have literary deconstructionists who argue that, "Paleoanthropological theories are origin stories, just like other origin stories. Many of them are 'just so' stories. And in the past, some of them had unpleasant racist implications."

So why study human fossils at all? What do they prove? How do you deal with those criticisms?

Johanson: You've really dissected out the three major lines of criticism. The first comes from the so-called "Scientific Creationists," although that's really an oxymoron. But we do get that type of criticism from those who believe in the revealed truth of the Bible. They do exactly that. They believe in the Bible. They accept it on faith. They make no distinction between what we know from facts and what we believe by faith. The creationist issue, which has been beaten like a dead horse, essentially boils down to the fact that science and religion are two entirely different ways of looking at the universe. One is based on the presupposition that you have to invoke some sort of supernatural intelligence, or being, or force. That presupposition is based solely on belief, rather than on fact; whereas science looks at evidence which is immediately and instantaneously available to anyone who has the open mind to examine it. One doesn't have to invoke any supernatural cause to explain, say, the Lucy skeleton. The skeleton is there. It's a fact. It's dated at 3.2 million years bp (before the present). It comes from a geological stratum of that antiquity and the skeleton is associated with the remains of extinct animals. We don't need to invoke any supernatural understanding simply to accept that as a fact.

An example I often use in my class is that we don't, in a religious sense, believe in gravity; we accept gravity. We don't evaluate gravity in the same framework that we evaluate our personally held religious or spiritual ideas. Gravity isn't moral or immoral. It's simply a fact. As I was sorting through the stuff in my office, I came across a story about the creationism taught in fundamentalist private schools in California. There on the front page of *The San Francisco Chronicle* of December 17, 1996, was a 5-by-7 photo of a sign on the desk of a 10th grade biology teacher that read, "Darwin Is Dead and He Ain't Coming Back!," which has now been turned into a bumper sticker. To which I responded in a letter to the editor of the newspaper, "Isn't it a shame that having nearly reached the millennium, we're still teaching ignorance to our students. Sir Isaac Newton is dead, but gravity ain't going away!" So maybe we can have that as a bumper sticker. We're just not going to change the minds of those who believe that the world is flat, that all the lunar landings were really staged in the Nevada desert and that we've never been in space. The only thing you can hope is that the United Airlines pilot who is flying you from San Francisco to London doesn't believe that the earth is flat, because he's never going to get you there.

Miele: Then let's consider the more intellectually serious criticisms. Your colleague Vincent Sarich, who is on our editorial board, and the late Allan

Wilson upset the whole paleontological apple cart back in 1967 with their molecular dating of human evolution.

Johanson: Vince is always a provocative guy—and a terrific guy!

Miele: When I was in graduate school, you had to either accept the paleontologists' view that the fossil record showed that the apes and humans split maybe 15 to 20 mybp, or you could go by the molecular evidence, which said the split was much more recent, maybe 5 mybp. Eventually, the Sarich and Wilson viewpoint won. The interpretation of the fossil record then had to be brought into accord with the molecular evidence. Is that how you see it?

Johanson: This issue of molecules and anatomy tells us a great deal about the branching patterns and the levels at which various species, or genera, or families, or other higher-level taxa, separated. As we compare the molecular data, and this has been done extensively with proteins, with genes themselves, and with serum albumin as Sarich did, we develop an understanding of the branching sequence for the evolutionary history of the particular taxonomic group we are looking at. That methodology can be traced back to Morris Goodman's work in the late 1950s and 1960s. He alerted us to the fact that those taxa that were more recently related to one another would not only show tremendous similarities in anatomy, and possibly behavior, but they should also show a great deal of similarity at the molecular level.

What Sarich and Wilson did was to go from Goodman's insight and assume that the molecular clock was moving along and changing and mutating at a fairly constant rate, from which we could then calculate dates of separation. Their initial estimate was that humans and African apes separated something like 5 mybp, which others have since been pushed back somewhat in time. But their date was nothing like the 15 to 20 million years that most people looking at the fossil record in the 1960s proposed. Their prime candidate for a protohominid in the fossil record at that date was *Ramapithecus*.

Miele: Which the critics delight in pointing out now turns out to be a fossil orang.

Johanson: Right. But if we go back and examine the original evidence that was used to establish the *Ramapithecus* scenario, we now see that it was a "just-so" story. I don't think anybody can deny that. From a jaw fragment, it was inferred that a delayed dental eruption sequence, which meant extended childhood, which meant significant parental care, which meant enlarged brains, which meant stone tools, which meant bipedalism. That scenario involved huge leaps of faith. Before we can reconstruct our evolutionary history accurately, we need to start with the basics—determining the geological context in which the fossil was found, providing a purely anatomical description of the fossil, presenting a metrical, anatomical comparison of that fossil with others. We can then put that information into a less subjective and more objective cladogram, in which we simply look at the shared, derived features and ignore the ancestral, conservative features. Then, after we have accumulated enough of that type of information, we can try to construct a phylogeny that incorporates time, dates of divergence, dates of extinction, and the degree of diversity within the fossil species. Only later, much later, can we intelligently script the scenario of human evolution. I don't think we've reached that point yet.

But we're always tempted to look for a quick fix. The climate changed, so bang, we evolved from *Australopithecus* to *Homo*. We don't know if that was really the case. More and more evidence is coming along which says that it was not.

Let me return to your specific question about anatomy versus molecules. Without a fossil record, it would be very easy to support the idea that a Miocene ape simply stepped into the "Magic Tunnel of Evolution" and emerged as an upright, White, European male, which is what you see in the pages of so many magazines.

Miele: It was in my high school textbooks.

Johanson: Without the fossil record, we wouldn't know that the tree of evolution was very bushy, that there were many branches, and most of them went extinct, with only one survivor. That's essentially what Darwin said back in 1859—that the tree of life was is in fact a bush, that it was not an inexorable march from a primitive to an advanced or derived creature.

Without the fossil record, you'd never know of the existence of *Australopithecus boisei*. You wouldn't even know about the robust australopithecine radiation. They represent an extraordinary adaptation, that we are now beginning to suspect underwent parallel evolution in southern and eastern Africa—separately. So in eastern Africa, you had *Australopithecus aethiopicus*, and then *Australopithecus boisei*. While in southern Africa, you had *A. africanus* and *A. robustus*. Yet the two groups converged on the same anatomical adaptations. That tells us something about the richness of the fossil record. Today, it's a very impoverished record. We're the only species of hominid alive. From that perspective, it's a very uninteresting world. Think of how much more interesting it would have been 35,000 years ago when you were marching across the European landscape and you said, "Who are those guys? They must be the Neandertals. We'd better stay away from them!" or whatever.

The fossil record is extraordinarily important for revealing to us the various branches of evolution that went extinct. They provide us with a remarkably poignant perspective of our place in nature. Even the general public that accepts the fact of evolution often thinks that we are the pinnacle of evolution and that we've been around forever and that we'll last forever. We've probably been around for maybe 150,000 to 200,000 years. But the fossil record reveals to us the mini-adaptive radiation of the robust australopithecines that probably lived across most of Africa, though the best places to find their remains today are eastern and southern Africa. They lasted for almost a million-and-a-half years. And yet they went extinct! So without the fossil record, we'd never know about the richness of our family tree. This is what makes the whole human story, the story of our place in nature, so rich and so attractive.

If we go back to the case of *Ramapithecus* and examine the original material, the original upper jaw that was used to reconstruct what was termed the "parabolic arch," upon which that whole evolutionary scenario was built, we find out that the midline of the jaw wasn't even present on that specimen. The reconstruction of the "parabolic arch" was a flight of fancy. From that basic mistake, an entire scenario was concocted, without necessary prerequisites of first developing a cladistic analysis, fleshing it out with a phylogeny, and then testing the various possible cladograms and phylogenies. Looking back, what we now see is a convergence of the fossil record and the molecular evidence, which alerted us to the fact that we are remarkably closely related to the African apes and couldn't have separated from them as early as 15 to 20 mypb, but rather more recently. But the real resolution of the *Ramapithecus* controversy came not from the molecular evidence. It came from the fossil evidence.

When we went back and looked at the Yale Peabody Museum specimen, YPM 13799 and saw that there was no midline on that specimen, so you could reconstruct either a human-like parabolic arch or an ape-like parallel tooth row, the fossil evidence suggested something was wrong. Then when *afarensis* was found in the 1970s and described, we found that it has a very parallel-sided, rectangular, ape-like dental arch, not a human-like parabolic arch.

The entire *Ramapithecus* issue was finally resolved when a more complete specimen was found in Pakistan by David Pilbeam's team. Pilbeam and Elwyn Simons had been the principal proponents of *Ramapithecus* as a protohominid. The skull of that specimen looks very, very much like a living orang. But again the fossils tell us something in terms of the mode of evolutionary change. While the narrow interorbital septum, its lack of frontal sinuses, the sort of ski-slope to the face, and many other features look very much like those of a modern orang, the post-cranial material suggests a more generalized anatomy that is very different from what we see in orangs today. This means that in terms of the sequence of evolutionary change, the skull obtained its modern form first and only later did the adaptations of the post-cranial skeleton, similar to those seen in the modern orang, take place. So here again the fossil record is very important because it tells us something about, going back to good old George Gaylord Simpson, not only the tempo, but the mode of evolutionary change—what happened first, what happened second—what was the sequence of events. Whether bipedalism came before big brains in the hominid line or big brains came before bipedalism was a huge issue that was not going to be settled in the test tube. Lucy and other fossils have given us the answer—we were upright long before our brains began to enlarge.

Miele: Let's take a quick look at that fossil with which your name will be forever linked. What does Lucy tell us about the evolution of our bodies and brains?

Johanson: In the book, we use Lucy as a metaphor for the whole species of *Australopithecus afarensis*. She tells us that at least as recently as 3.2 million years ago, the australopithecines were still quite ape-like in their general, overall anatomy. They had fairly long arms compared to their legs, which is probably evolutionary baggage left over from a more quadrupedal, arboreal, climbing ancestry, like we see in chimps today. We see that her pelvis has a very small pelvic outlet, which means that the brains of the newborns were very small. The adult brains were not anywhere near the size of modern humans. Very little of her braincase has been preserved, but fortunately we do have enough of the occipital (the back of the skull), to tell us that she might have had a braincase that was well under 400 cubic centimeters. Yet looking in detail at the pelvis, the knee, and the ankle, there's no question that the primary locomotor adaptation was different from that of a quadruped. Lucy was a biped, maybe not identical to us today in terms of locomotor mode, but one that had definitely made a significant shift in locomotor repertoire from quadrupedalism to bipedalism. The shape of her thorax is more of a pyramid (again, a primitive feature), rather than a broad barrel chest, seen in modern humans. Overall, Lucy presents an interesting amalgam of ape-like characteristics and more derived features, such as bipedalism. (See Figures 8.1, 8.2, and 8.3).

Miele: A lot of evolutionists have fixed their eyes on the canine teeth.

Johanson: Canine teeth are very interesting and at the moment we have a somewhat enlarged sample we're beginning to examine. There does not appear to be a very major difference in canine size between males and females in *afarensis*, as there would be in chimps or gorillas. Both male and female *afarensis* had feminized (i.e., reduced) canines. On the other hand, there is a tremendous degree of sexual dimorphism in body size. The males were much larger than Lucy. Many people wondered when Lucy was first found whether this was some sort of aberrant fossil, maybe that of a dwarf. No, it wasn't, because there's nothing abnormal about its anatomy to indicate that. So the question then arose, does she represent a different species from the larger specimens found at Hadar (Ethiopia), a hypothesis we initially entertained ourselves.

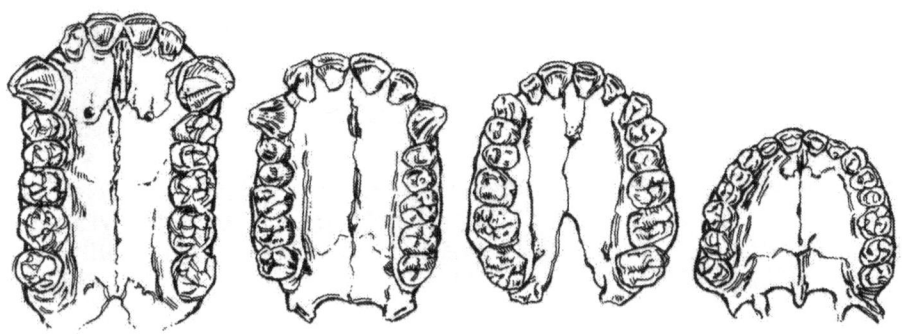

Figure 8.1: Shape of Upper Jaw and Dental Arch and size of Canine teeth in Gorilla, Chimpanzee, *Australopithecus afarensis* ('Lucy') and Human. Note that like humans, Lucy has much smaller canine teeth than the two ape species, who have large canines and parallel tooth rows. [Adapted from Howells, Fig. 27, p. 81 and Johanson and Edey, p. 265].

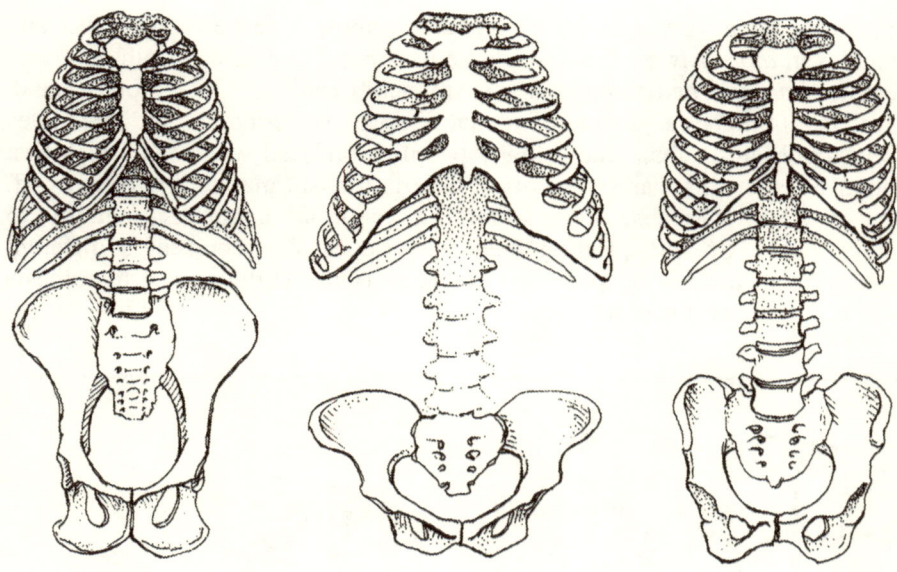

Figure 8.2: Rib cage and pelvis in Chimpanzee, *Australopithecus afarencis* ("Lucy") and Huamn. Lucy's rib cage is similar to the pyramidal rib cage of the chimp, while the pelvis is intermediate, with a small pelvic outlet. [Adapted from Howells, Fig. 25, p. 79].

Figure 8.3b: Chimpanzee canines are larger in proportion to the rest of their teeth than either humans or *Australopithecus afarensis.* [Adapted from Johanson and Edey, p. 265].

Figure 8.3a: Chimpanzees show much greater sexual dimorphism in canine tooth sieze (M>F) than either humans or *Australopithecus afarensis.*

First, let me revisit your question about the creationists who say, "Well one year, the evolutionists say this, and next year they say that." That's one of the charms of doing science. That's one of the wonderful things about doing science. You go into the laboratory or out into the field and find something new, and all of a sudden, you have to change your mind. That just means you have a scientific approach, you have an open, inquiring mind. You have a skeptical mind that says, "I don't want to put all my eggs in this basket. I want to adopt the strategy more typical of an ostrich, who lays her eggs in different nests, so if lions destroy one nest, all her genes aren't gone because she's got eggs in other nests." We want to entertain a series of hypotheses. Even if we become very committed to one them, something may come along and enlighten us to change our minds. That's one of the great rewards of participating in science. The more evidence that comes in, the more exciting it is for entertaining very different opportunities and possibilities. Not only is this not a criticism of science, it's one of its strengths—we can change our minds when new evidence comes to light and causes us to radically revise our way of looking at things.

Returning to sexual dimorphism, when we look at *afarensis*, we find a considerable amount of it. Lucy was 3-1/2 feet tall, males were probably about 5 feet tall; Lucy was only 50 to 55 lbs., males 100 to 110 lbs. Males were almost twice the size of the females. This has been the focus of a number of scholars, because of Owen Lovejoy's hypothesis. (Lovejoy, 1993) Seeing that there was little difference between the canine teeth of males and females, he took that as evidence that males didn't need large canines to compete with other males for access to estrous females, and concluded that our ancestors were monogamous.

Miele: How do you evaluate Lovejoy's hypothesis against the evidence that most contemporary societies are not monogamous while the great apes are sexually dimorphic and show a variety of mating patterns.

Johanson: That's exactly the point. I prefer not to look at modern humans, because we're so derived and depend so much on culture that almost anything is possible. But when you look at a creature like Lucy who was, I'll use the term, more in tune with the natural world in which she lived, probably a great deal of her anatomy reflects the world she lived in. There is a considerable amount of criticism of the Lovejoy model, which is emerging more and more. A recent book on women in paleoanthropology by Laurie Hager (1997) includes substantial criticism by many primatologists and behavioral researchers who look at the size differences between males and females in *afarensis* and point out that monogamy is not the pattern we see in contemporary primate societies that have comparable degrees of sexual dimorphism. Instead, we see harem groups in gelada baboons or in gorillas, or we see polygynous groups, or we see multi-male—multi-female groups, where there's a mixture of mating combinations. So that it was a great leap of faith to go from just canine size to reduced aggression to monogamy.

Maybe there was reduced aggression in *afarensis*. But look at bonobos, where you have a significant male-female difference in canine size but a female-dominated society in which conflict is resolved by sexual behavior. This is totally different from what you would conclude simply by extrapolating from the relative tooth sizes. That's the danger of constructing evolutionary scenarios based on just one or a small number of features found in the fossil record.

Miele: In addition to sexual dimorphism, your book also illustrates the racial or populational variation in living humans. We also see that individuals change during development from childhood to adulthood, and we don't even know whether the trajectory of that change was the same in the fossils as it is today. So, being a good skeptic, when you examine a family tree, such as the one by Bernard Wood which you have adapted in your book (See Figure 8.4b), how do you decide whether what you're seeing in the fossil record is a real evolutionary difference, versus a difference between sexes, or a racial/populational difference, or a maturational difference?

Johanson: Almost every feature you've mentioned, Frank, was addressed many years ago by Wilfred Le Gros Clark. He was extremely observant in his writing [see *The Antecedents of Man: An Introduction to the Evolution of Primates*, 1959] on the fossil evidence for human evolution, where he said you cannot take the Taung baby (an australopithecine fossil) and compare it to a grown-up chimpanzee, or compare a baby chimp to a fully grown human. You have to be aware of the ontogenetic age of your fossils and compare babies with babies and adults with adults. You also have to be well aware of pathologies, because some nutritional deficit may have influenced the shape of the teeth or the jaw. Males should be compared against males and females against females.

People sometimes ask why we have to go out and keep finding more fossils. As we build a collection of specimens, we find out about the range of variation within a species. If we look at a large collection of skeletons of chimpanzees or gorillas, we can use them as a baseline that tells us the variation in size and morphology within a single species. We can then use that statistic as a yardstick to evaluate what we find in the fossils.

When we first look at the very large variation in size in a sample like the *afarensis* material at Hadar, we immediately consider a couple of hypotheses. Hypothesis #1 is that the size differences are due to the fact that there are really two different species at the site. Hypothesis #2 is that there's only one sexually dimorphic species. To test those hypotheses, we examine more carefully the range of variation in tooth size, or jaw size, or the morphology and anatomy of the face. In the case of *afarensis*, we found that there was quite a difference in size, but it was comparable to what you find between male and female gorillas or chimps. It's just what's called an allometric difference, a scaling difference, where the features that identify the species are the same in both large and in small individuals.

That adds more strength to Hypothesis #2, that there was only one species at Hadar between 3 and 4 million years ago. But it's still only a hypothesis. There may be another species there. At the moment, however, most of the evidence suggests that there was only a single species. In the future we may find that there was significantly more diversity, but what we see now is consistent with the range of variation we see within a single species in living chimpanzees, or in gorillas, or even in collections of other fossil species.

This brings me back to your question about how skeptically we should look at the family tree shown on Figures 8.4a and 8.4b. What stands out most to me in that representation is the period between four to three mybp. Look, there's only one species there. And that, to me, immediately raises a red flag. Are we missing a species somewhere? Was there another species living in a somewhat different habitat that we haven't discovered yet? Why did hominids become so speciose (i.e., species plentiful) after 3 mybp but so conservative between 4 and 3 million years? That is an issue. And I leave that more as a question, a source of inquiry, a stimulus to future discoveries and interpretations, because if you look at the period just before 4 mybp, it looks like there were two different taxa. So now, some scholars are suggesting that *Australopithecus anamensis* may have overlapped in time with *Ardipithecus ramidus*.

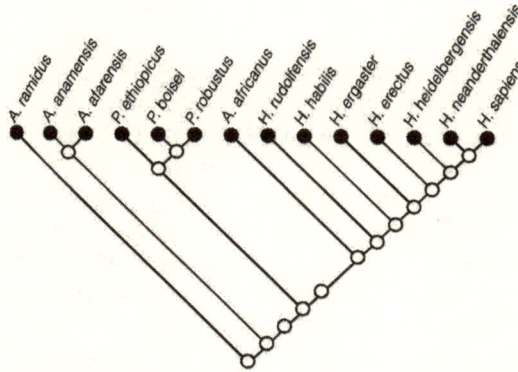

Figure 8.4a: Hominid Cladogram–a branching diagram that shows shared anatomical features between fossils and living hominids, and sorts them into related groups. [Adapted from Johanson and Edgar, p. 38, after Tattersall].

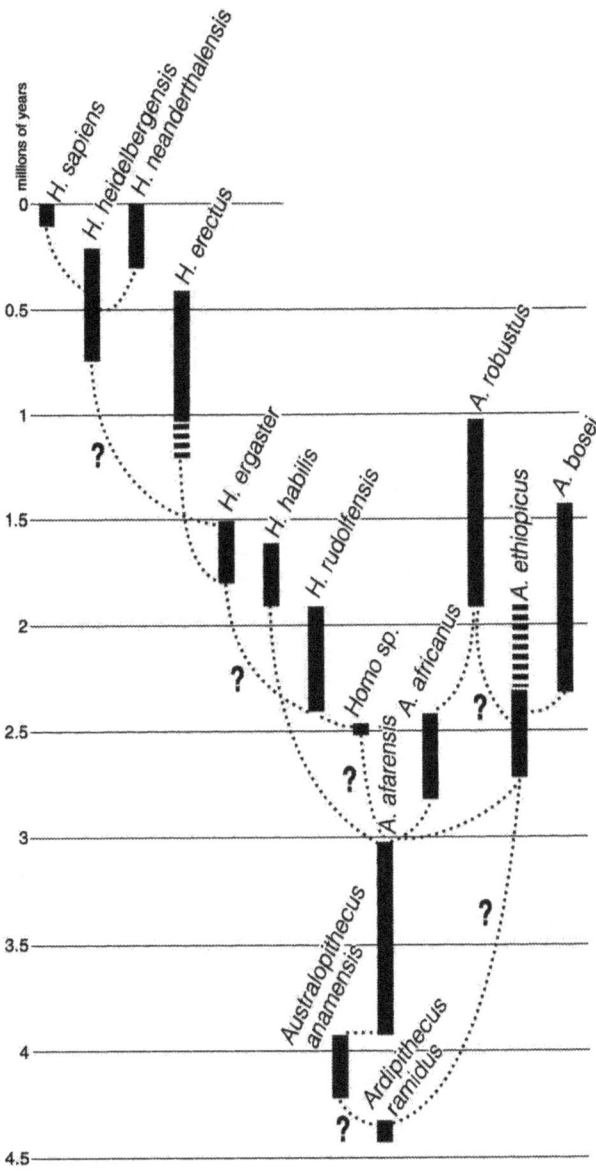

Figure 8.4b: A Phylogenetic Tree adds the dimension of time and proposes explicit evolutionary lineages. Note how only one species occupies the 4-3 mybp period, suggesting there are more species yet to be found. [Adapted from Johanson and Edgar, p. 38, after Wood]

Miele: Following up on the question of how you interpret the fossil record and test competing hypotheses, Carleton Coon's book, *The Origin of Races*, presented his controversial view of human evolution, especially as regards the parallel evolution of the major human races. Current interpretations of the fossil record certainly do not support his view, which was controversial even at the time. However, his book presented a wealth of data with which others could test his hypothesis or develop and test hypotheses of their own. And the subject of the origin of racial or populational differences remains unresolved and controversial.

Johanson: When we were graduate students at the University of Chicago we read *The Origin of Races* and we were appalled by the misuse of the book. It was packed with data and information. And it was the first time that a book with that information was made readily available. We tended to ignore all of the open nerves that were touched by misinterpretations of the book and realized how vital and important that book was.

Miele: In your book you have a photograph of the Kow Swamp skull from Australia. It is a *Homo sapiens* skull from only about 10,000 years ago that has a much lower cranial capacity than the Neandertals. Kow Swamp also shows a number of conservative, ancestral features, such the brow ridge and a sloping forehead. You also can see those features in some living humans today. And we also see a broad range of cranial capacities in living humans, all of whom live in the modern technological world and lead thoroughly modern lives. So either the anatomical differences in living humans, either within groups or the average difference between groups, must mean something behaviorally, or how do you know those differences meant something when seen in the evolutionary record? If someone suddenly cloned a Neandertal, or a *Homo erectus*, or an *Australopithecus afarensis*, how do we know that they would be behaviorally different from people alive today?

Johanson: You've really sent up a balloon here. There's so much to consider. Let me go back to some very basic stuff about the *Homo sapiens* skulls we have illustrated. We've just chosen at random a group of skulls and illustrated them to show that there is a significant range of variation in modern humans. And there are clearly populational differences. Now, the population differences that we see between, say, people who live in Lapland and people who live in Australia are clearly discernible. But when we look at them in the overall behavioral context, we find out that whether you are a native Australian or a native Laplander, what's going on inside the braincase is essentially the same. They're capable of all the same sorts of behaviors as any living member of *Homo sapiens*.

We don't always have good answers to the question of why groups of living humans look different from one another. We can look at Eskimos, who live in very cold areas and tend to be cold adapted, and see that they have short, squat bodies which tend to conserve heat. As did Neandertals. So in both cases, body shape probably represents a very consistent micro-adaptation to the environment. On the other hand, why are there other populational differences today? I think that boils down to the fact that most Laplanders marry Laplanders, and most native Australians marry other native Australians. Generally, we tend to marry people who look very much like ourselves, our parents, our siblings. Until relatively recently, there was not much opportunity for as much gene flow as there is today. That's probably why we developed these different morphotypes, these different skull shapes. Some people now think that the shape of Kow Swamp skull, for example, also reflects a distortion due to cultural practices such as binding the skull.

Miele: There are some anthropologists, however, like Milford Wolpoff and Alan Thorne, who look at the brow ridges in the Australian samples and say that they reflect the continuity of, if not races as Carleton Coon believed, at least those morphological features. And some, like Eric Trinkhaus, argue that the pronounced nasal morphology of living European populations reflects a continuity from the Neandertals, while Out-of-Africa proponents like Christopher Stringer disagree.

Johanson: The issue of continuity has not been thoroughly resolved. I lean more toward the Out-of-Africa hypothesis, because the earliest sapiens that we have in the Middle East, like the Qafzeh specimen, show a more tropical, African type of build. And the earliest evidence we now have for anatomically modern humans comes from Israel and the Horn of Africa. So I believe *Homo sapiens* originated somewhere in that African arena. But I'm not totally closed to the idea that it could have happened in another place. Alan Mann at University of Pennsylvania and I have spoken about this a number times and there is a possibility of early sapiens in China. And if that is the case, maybe *Homo sapiens* first arose there.

Miele: Can the molecular evidence resolve the issue? I read recently that analysis of some Neandertal DNA seemed to rule out the possibility that they contributed to the modern human gene pool (Kings, et al. 1997). Can you triangulate between the fossil record and the DNA data to test the various hypotheses?

Johanson: Yes. We need to look at the totality of evidence. When we develop a cladogram for our fossils, we obviously rely upon comparative skeletal anatomy. But when we deal with living species, we can use behavior, physiology, as well as the molecular data. In the past it wasn't possible to get molecular data from the fossil record. I find the Neandertal DNA evidence extraordinarily fascinating. I would be inclined to grab on to it as providing the ultimate proof that Neandertals did not leave genes in modern human populations. But it is, after all, the DNA of one individual. It is only mitochondrial DNA, not nuclear DNA. So it's another piece of evidence, but I don't think it slams that door completely shut. I think we're going to need a lot more DNA before we can make a really strong call. The people who promote the Out-of-Africa model have glommed on to the Neandertal mitochondrial DNA as the final nail in the coffin of the Multiregional model.

Miele: Some of them also glommed on to the argument of anatomist Philip Lieberman that his reconstruction of the anatomy of the Neandertal larynx ruled out the possibility of speech. But as you point out in the book, that one didn't hold up.

Johanson: Exactly. The hyoid bone found in the Kebara 2 specimen is virtually identical to that of a modern human, which may imply Neandertals were capable of speech. It's going to be very difficult, if not impossible, to determine the origin of language, which is an extraordinary adaptation. What's going on in language is going on inside our brain and how much do we really know about how the brain works? I'm fairly convinced that Chomsky and the others, who today are considered the bad guys, are right in their view that the ability to speak is hardwired into our brains. No matter which society or culture you grow up in, you speak that language perfectly, because the basic underlying structure necessary to learn language is already plugged into your brain.

Miele: Any guess as to when that module got slotted into the brain?

Johanson: I tend to favor a later arrival. Others like Phillip Tobias favor a younger arrival. Human language is part of a continuum of communication that ranges from body posture and hand motion to articulated speech. I think that modern syntactic language, as we are using it right now to symbolically represent ideas, emotions, whatever, is fairly late and associated with the Upper Paleolithic creative explosion. I think Neandertals were pretty good at what they did and undoubtedly had pretty good language, but they didn't have a lot of planning depth. They didn't have a way of symbolically representing the world around them to the degree that modern *Homo* *sapiens*, the Cro-Magnons who left the cave paintings did. And at the moment, that's the cultural, physical, artefactual evidence we can point to that shows a fairly rapid explosion occurred in the Upper Paleolithic. I see it as providing hard evidence that the Upper Paleolithic people were no longer living in a dark, murky world, but beginning to celebrate the world in which they lived.

Miele: Any speculation as to what, if it wasn't God zapping them, put that extra module in their brains?

Johanson: You know, I have no idea. I'm looking at my Macintosh sitting on my desk. Until you put in the software, it's like the paleolithic brain. It's essentially dead. But once you put in that software and communicate with the machine, and as it uses that software to perform certain tasks, it can perform all these wonderful functions. How that software got plugged into our brains 50 or 100,000 years ago, I just don't know.

Miele: Scholars have looked to things such as climate change, you've mentioned Lovejoy's hypothesis, Dean Falk has proposed the Radiator Hypothesis (i.e., that bipedalism evolved so we could dissipate heat more efficiently), to explain the changes we see in the fossil record. There's always this search for some external driver—if not God, then something in the environment. To what extent does culture drive itself? To what extent have we shaped our own evolution, once we started running that cultural software?

Johanson: That's an extremely powerful and useful way to look at human evolution. We tend to be very adaptable creatures. We can adapt to virtually any and every environment through cultural means. We also tend to be very adaptable in terms of our physiology. We're not highly adapted creatures. My favorite example that I use in my classes is Gary Larson's cartoon of the cows watching TV when the phone rings. One cow turns to the others and says, "There it goes again. And we just sit here without opposable thumbs."

Human evolution is not driven strictly by biology. It is not driven strictly by culture. It is really a coevolution of biology and culture. There's a section in the book entitled, "Is Human Evolution Different?" It is different, because culture is a very major element in our evolutionary success. It is possible, as E. O. Wilson and Sherwood Washburn have argued, that culture takes on a very important role in shaping the species itself. When we look at the arrival of anatomically modern humans, it's not just biology that's driving their evolution. It's also culture.

I'm sure you're aware of the huge controversy in the anthropology department at Stanford University that was highlighted in *Science* a few months ago (Gibbons, 1997). There are some anthropologists who believe that when culture comes in the door, biology goes out the window. They believe that we were born with our minds a Lockean tabula rasa.

Miele: That the software runs with no hardware platform underneath it?

Johanson: Yes, that's what they're saying. But given the millions of years of natural selection that have shaped us, both biologically and behaviorally, there has to be an important element in our behavior that is related to the genes and based in biology, as E. O. Wilson and the sociobiologists and evolutionary psychologists have argued. So to really understand our evolution, we cannot look for the "magic bullet" that made us human, or assume that culture sprang up spontaneously. Human evolution took place through a series of events that ultimately expressed themselves in the interaction of biology and culture.

Miele: Finally, where in time and space do you think the next major advance in our knowledge of the human fossil record will take place? Will it be in the early part of the record, around the time where hominids and pongids split, or much later on when modern *Homo sapiens* appears and the racial or populational variation begins? Will it come from discoveries outside of Africa?

Johanson: One area that I think is very, very exciting is the study of the deposits that date earlier than 4 mypb. Tim White's work on *Ardipithecus ramidus* has opened that window for us. If we go back to the Miocene period, we see an incredible diversity of ape species. They occupied Europe and Asia, as well as Africa. David Pilbeam has written a very interesting and provocative article (Pilbeam, 1996) that puts forth a new view of what protohominid looked like. He argues that it would be very chimp-like, with thin dental enamel, which is very different from the prevailing view that it would have thick (i.e., human-like) dental enamel. But in fact, we don't have a really good, solid, protohominid ancestor from the Miocene. We need to go out and find one.

The other area I find intriguing is the transition from *Australopithecus* to *Homo*. Since the recent discovery at Hadar of the oldest well-dated evidence for the genus *Homo*, I'm more and more interested in reconstructing that link, which, I think, represents an extraordinarily important moment in human evolution. We should abandon trying to think up grand scenarios of global change and environmental desiccation to explain that transition. Instead, we have to go out and find the real evidence of the transitional forms between *Australopithecus* and early *Homo*. First we need to understand what those differences were and how they were expressed. Then we can build a cladogram, and after that develop a phylogeny. Only then should we try to paint the whole big picture. We may find out that, just as there were many different species of *Australopithecus*, there will also be many different species of *Homo*. We're beginning to get an insight into that looking at *Homo habilis*, *Homo ergaster*, and *Homo erectus*. It may be that the tree of evolution was very speciose (species plentiful) at that time, but that only one branch of tree survived to evolve into ourselves.

Miele: Thank you for talking with *Skeptic* and for making much clearer the complex story of our evolutionary history.

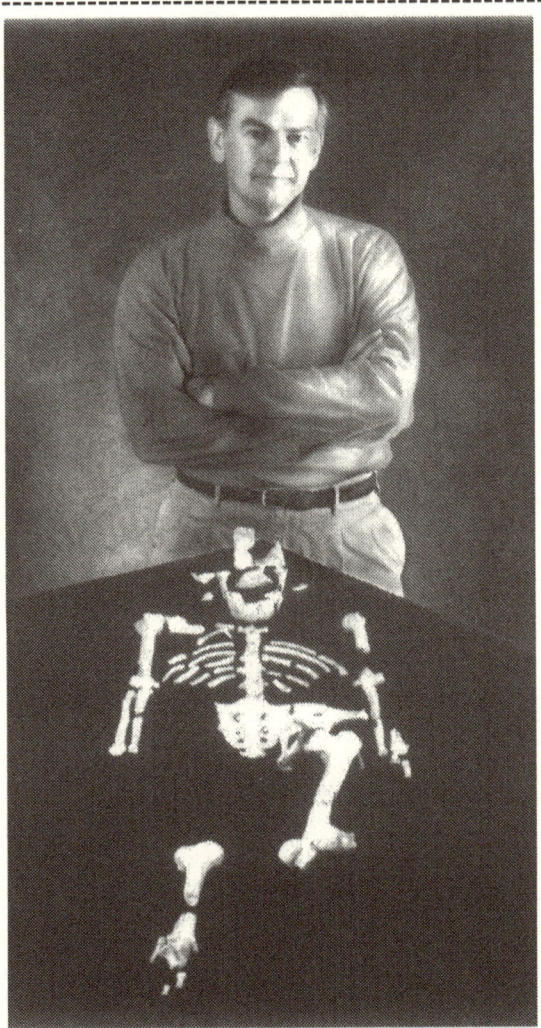

Figure 8.5: Donald Johanson with "Lucy"

Frank Miele

References

Asimov, N. 1996. "A Look Inside and Evangelical Classroom," *San Francisco Chronicle*. 17 December 1996, p. A-1.

Coon, C. S. 1962. *The Origin of Races*. New York: Knopf.

Falk, D. 1992. *Braindance: New Discoveries About Human Origins and Brain Evolution*. New York: Holt.

Gibbons, A. 1997. "Cultural Divide at Stanford," *Science*. 20 June pp. 1783-1784.

Hager, L. (ed.) 1997. *Women in Human Evolution*. New York: Routledge.

Johanson, D., and Edey, M. 1981. *Lucy: The Beginnings of Humankind*. New York: Simon and Schuster.

Johanson, D., and Edgar, B.1997. *From Lucy to Language*. New York: Simon and Schuster.

Krings, M., Stone, K A. Schmitz, R. W., Krainitzki, H., Stoneking, M., & Paabo, S. 1997. "Neandertal DNA sequences and the origin of modern humans." *Cell*, 90: 19-30.

Le Gros Clark, W. 1963. *Antecedents of Man*. New York: Harper.

Lovejoy, O. 1993. "Modeling Human Origins: Are We Sexy Because We're Smart, or Smart Because We're Sexy?" pp. 1-29 in D. T. Rasmussen (ed.). *The Origin and Evolution of Humans and Humaness*. Sudbury, MA: Jones and Bartlett.

Pilbeam, D. 1996. "Genetic and Morphological Records of the Hominoidea and Hominid Origins: A Synthesis." *Molecular Phylogenetics and Evolution*. Vol. 5, No. 1, pp. 155-168.

Simpson, G., 1949. *Tempo and Mode in Evolution*. New York: Columbia University Press.

Stringer, C., and McVie, C. 1997. *African Exodus: The Origins of Modern Humans*. New York: Holt.

Tattersall, I. 1995. *The Fossil Trail*. New York: Oxford.

Trinkhaus, E. and Shipman, P. 1993. *The Neandertals: Changing the Image of Mankind*. New York: Knopf.

Wolpoff, M. and Caspari, R. 1997. *Race and Human Evolution: A Fatal Attraction*. New York: Simon and Schuster.

Wood, B. 1994. "The Oldest Hominid Yet" *Nature*, 371:280-281.

CHAPTER 9

A QUICK & DIRTY GUIDE TO CHAOS AND COMPLEXITY THEORY

THREE RACE HORSES & FOUR HOBBY HORSES

BOOKIES HAVE A SAYING, "There are three horses who have never come in win, place, or even show. Their names are Coulda, Woulda, and Shoulda." Bookies live—and sometimes die—by the accuracy or inaccuracy of their predictions. Their performance is immediately obvious to all involved.

The same, however, cannot be said of many academics, even less so of pop-science writers. Despite what you've heard about the hypothetico-deductive method being the touchstone of science, it is the exception when predictions by academics are put to the empirical test and they have to live or die by the results. It's all too easy for them to come up with 'supplementary hypotheses' to explain their way out. (Try out a supplementary hypothesis on Tony the Crippler when he knocks on your door to collect the two grand you dropped on a pony that finished dead last. You'll soon drop dead). This lack of accountability is further parlayed as one moves from the physical sciences, to the biological, to the behavioral, to the social sciences, reaching the level of a dead cert in literary studies.

The worst such abuses occur when terms that have a clearly defined meaning, usually mathematical, in the physical sciences are imported into literary studies as metaphors. By the time they reach pop books, a good skeptic's baloney detector should be red-lining. Latest to make this transition are the two C-words—Chaos and Complexity, and their hybrid offspring, Contingency and Counterfactuals. Together they are the academic counterparts of Coulda, Shoulda, and Woulda. They are the four hobby horses of Chaostory. What makes this jockeying all the easier is that the restricted, stipulative definitions of these terms, if not the opposite, are certainly a long way from our understanding of them in everyday discourse.

CHAOS: IT'S ANYTHING BUT HELTER SKELTER.

Let's start with chaos. The dictionary defines it as: "(1) utter confusion or disorder; (2) the formless matter supposed to have preceded the existence of the universe." *Roget's Thesaurus* gives "disorder, derangement, and anarchy" as synonyms, and "order, uniformity, regularity, and symmetry" as antonyms. But in the world of Chaos/Complexity theory one encounters such terms as "deterministic chaos" which results from "deterministic dynamical equations." Conceptually, we are told, "chaos is intrinsic to the system and clearly distinguished from the effects of random or 'stochastic' fluctuations in the external environment." Therefore, distinguishing deterministic chaos from stochastic ('true') chaos "is one of the principal hurdles that confronts 'chaologists'—scientists working with potentially chaotic systems." (Conveney and Highfiled, 1995, p. 174). What is the difference between deterministic chaos and stochastic ('true') chaos? The answer is the Attractor, which can be either a Fixed Point Attractor or a Strange Attractor. This all really makes sense within the world of physics and non-linear mathematics, but becomes meaningless chaobabble when applied elsewhere.

COMPLEXITY: THE SEARCH FOR SIMPLICITY

My dictionary defines complexity as "intricate, knotty, or perplexing." Its use in Chaos/Complexity theory certainly is perplexing, because there it means a search for simple rules that can explain how the universe can "start with a few types of elementary particles at the big bang, and end up with life, history, economics, and literature" (Bak, p. 2). How can this happen? The answer offered is self-organized criticality ["criticality," another C-word], "the tendency of large systems with many components to evolve into a poised, 'critical' state, way out of balance, where minor disturbances may lead to events called avalanches. Most changes take place through catastrophic ["catastrophic" –yet another C word] events rather than by following a smooth gradual pat" (Bak, 1996, p.2).

If the butterfly flapping its wings in Brazil and setting off a tornado in Texas has become the mantra of Chaos theory (even though there's no evidence that a gaggle of butterflies has ever so much as generated a zephyr), Per Bak's Child's Sand Pile is the icon of Complexity theory (Figure 9.1). He describes it as follows:

> In the beginning, the pile is flat, and the individual grains remain close to where they land. Their motion can be understood in terms of their physical properties. As the process continues, the pile becomes steeper, and there will

be little sand slides. As times goes on, the sand slides become bigger and bigger. Eventually, some of the sand slides may even span all or most of the pile. At that point, the system is far out of balance, and its behavior can no longer be understood in terms of the behavior of the individual grains. The avalanches form a dynamic of their own which can be understood only from a holistic description of the properties of the entire pile rather than from a reductionist description of individual grains: the sand pile is a complex system. (Bak, p. 2)

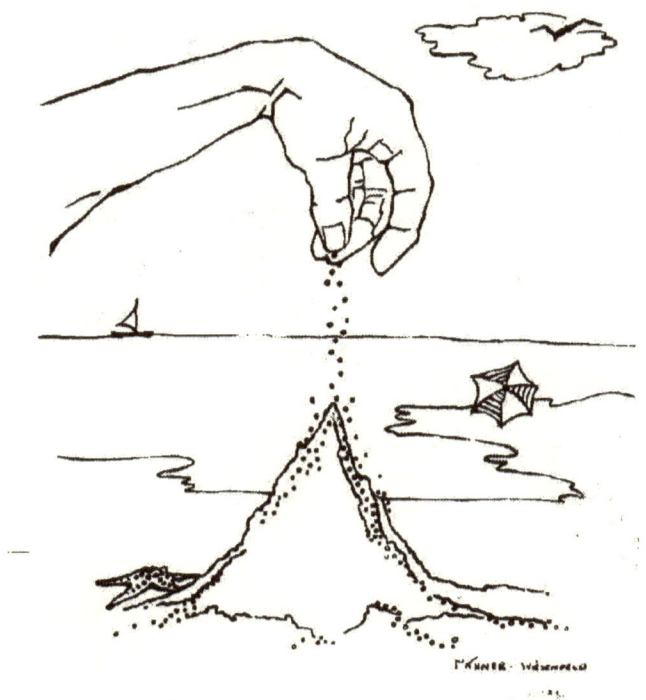

Figure 9.1: The Icon of Complexity Theory: Per Bak's Sand Pile. When a system (i.e., the pile) reaches a critical point and an avalanche occurs, its behavior is the result of the dynamics of the overall system, not of each individual element (grain of sand).

So taken by the sand pile metaphor was then Senator Al Gore that he cited it in his book, *Earth in Balance,* and credited it with transforming his view of the world:

> The sand pile theory—self-organized criticality—is irresistible as a metaphor; one can begin by applying it to the developmental stages of human life. The formation of identity is akin to the formation of the sand pile, with each person being unique and thus affected by events differently. A personality reaches the critical states once the basic contours of its distinctive shape are revealed; then the impact of each new experience reverberates throughout the whole person, both directly, at the time it occurs, and indirectly, by setting the stage for future change…. One reason I am drawn to this theory is that it has helped me understand the change in my own life. (Bak, p. 63)

Bak shows that such seemingly unrelated phenomena as Raup's data on extinctions, Mandelbrot's analysis of commodity prices, and Gutenberg-Richter's on earthquake magnitude all can be mapped to similar graphs. (Figure 9.2). In each graph we see that the number of incidents decreases as the magnitude of the event increases. Bak then argues that this is evidence that we should look for underlying general laws to explain all such behavior, rather than turning to narratives to describe each incident in terms of its unique precipitating characteristics (such as bad news from the Fed or an accumulation of stress in the earth's crust).

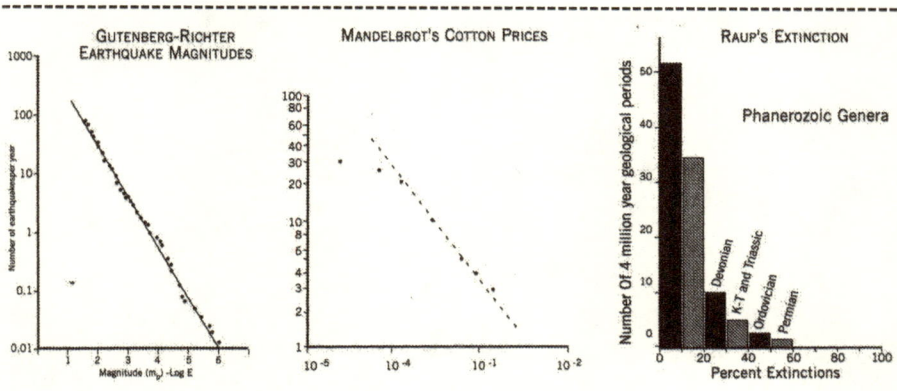

Figure 9.2: Three Examples of Seemingly Unrelated Phenomena that Follow (or at least can be seen to follow) a Common Pattern: Guttenberg-

Richter earthquake magnitudes, Mandelbrot's cotton prices, and Raup's extinction data. The simplest explanation is simply that big events are rare.

But the net worth of Bill Gates and any of 39 other internet tycoons exceeds that of the majority of the earth's inhabitants combined or the Gross National Product of some nations. Is the saying "the rich get richer while the poor get poorer" a fundamental law of nature not to be tampered with? It is unlikely many advocates of extending chaos and complexity theory beyond the physical sciences would buy into that one.

CONTINGENCY AS DISINGENUOUS DETERMINISM

The dictionary defines contingent as "1. dependent on something not yet certain. 2. accidental or fortuitous" and "contingency" as "dependent on chance, an uncertain event." In everyday discourse the term is associated with the fee arrangements made between lawyers who wait outside hospital emergency rooms for potential clients for whom they can sue the state or a large corporation, hoping they'll collect "walking away money" to settle a "slip and fall" liability case out of court. For present purposes, however, these terms owe their origin to Stephen Jay Gould. The sacred scripture for the doctrine of contingency is his book *Wonderful Life*, with the proof texts appearing on pages 48 and 287:

> I call this experiment "replaying life's tape." You press the rewind button and, making sure you thoroughly erase everything that actually happened, go back to any time and place in the past..... Then let the tape run again and see if the repetition looks at all like the original.
> ...[the] basic principle of contingency—a replay of the tape yielding an entirely different but equally sensible outcome; ...small and apparently insignificant changes ... lead to cascades of accumulating differences. (Gould, 1989, p. 63, p. 48).

The theme of *Wonderful Life* is to dispel the notion that we (i.e., *Homo sapiens*) are the inevitable product of progressive evolution. Gould argues that of the myriad of wild and woolly creatures alive in the Cambrian seas, it was only a matter of bad luck, not bad genes, that so many disappeared, leaving the field open for the evolution of our line.

But how meaningful a concept is contingency and how telling is Gould's analysis? At best, contingency is a non-theory in that it opts for the importance of freak events over general processes. Cosmology, evolution, or history become "just one damned thing after another." At worst, rather than an alternative to determinism, contingency is disingenuous determinism because it argues that the freak event which wiped out one line of evolutionary development foreclosed all the others and forced the remaining one on a rigidly determined course.

Granted, if our great ancestors hadn't drawn the lucky hand and survived the great Cambrian extinction, we wouldn't be here and you wouldn't be reading this article. But can we really be sure it was just a matter of luck rather than some characteristic such as adaptability that led to their survival? Also, the evidence on which Gould hangs his case—the alleged proliferation of 'weird' life forms in the Cambrian—has been subject to criticism. Some of these wildly weird creatures may be the result of reconstructing them upside down. A rightside up orientation produced very normal looking specimens. (Cohen and Stewart, 1994, p. 132) Even if they did exist in their woollier reconstructions, this is not proof that evolution would not have thrown forth and selected for other life forms possessing certain identifying characteristics of ours and other species. Repeatedly throughout the geological record we see different lineages successively evolving similar anatomies and behaviors in order to exploit available niches. Several independent evolutionary lineages (insects, bats, birds, and pterosaurs to name a few) have developed such complex processes as flight, the eye (chordates and cephalopods), or a brain capable of solving some form of logical problem (apes, elephants, and at a simple enough level, even the octopus). Though separated by 80 million years of evolutionary history since their last common ancestor, the Tasmanian (marsupial) wolf and the true wolf looked and lived a lot alike. The Tasmanian wolf actually survived up until 1933, so we have reports of its natural behavior and even films of it in captivity. As theoretical biologist Brian Goodwin explained:

> Suppose you reran the Big Bang. What are the chances of getting the same periodic table of natural elements, the same ninety-two combinations of protons, neutrons, and electrons? Pretty good, or so I'm led to believe. I think of a rerun of the Cambrian explosion in the same way, not to the same extent perhaps, but as an image.... a rerun of the Cambrian explosion would produce a world much more like the one we know than Steve Gould says. It wouldn't be identical to the one we know, but there may be a lot of similarities, ghosts we'd instantly recognize. (Quoted in Lewin, 1992, p. 74)

When the Chicxulub asteroid wiped out the dinosaurs, birds and land and marine mammals arose to exploit similar niches. And if, playing the contingency card, Chicxulub had missed the earth, it's a better than even bet that some line of the dinosaur tree would have evolved increasing ability to extract, process, and use information from their environment (i.e., intelligence) similar to what the fossil record shows took place for mammals and especially primates. (Russell, 1992).

We can see this clearly in Harry Jerison's data on the evolution of brain and intelligence. (Jerison, 1972) Figure 9.3 shows a graph of the brain weight/volume to body weight/volume for living mammals, fossil mammalian ungulates, and for living and fossil reptiles. The polygon formed by the brain-to-body ratio of the living mammals is distinct from that for the living and fossil reptiles. To the extent that brain-to-body ratio measures information-processing ability, these data show a progressive trend in evolution. The gray zone of archaic ungulates shows that they fall outside the range of living mammals, but are closer to them than to either the living or the fossil reptiles. This is evidence of progress across phylogenetic lines.

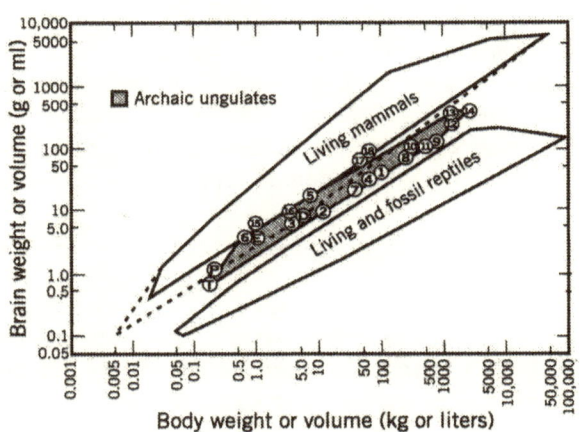

Figure 9.3: Evidence for Progress in Neural Complexity over Evolutionary Time: Information processing ability, as measured by the ratio of brain size to body size, is greater in mammals than in reptiles, with extinct mammals falling in between the two groups.

How did such progress take place? Chaos, complexity, contingency? Figure 9.4 depicts a set of bell curves that provide the answer—adaptation brought about by an arms race. Note first that for each era in geological time the predators (carnivores) have larger relative brain-to-body ratios than their prey (ungulates). Then note how the bell curve for the predators pulls the bell curve for their prey to successively higher mean brain-to-body ratios.

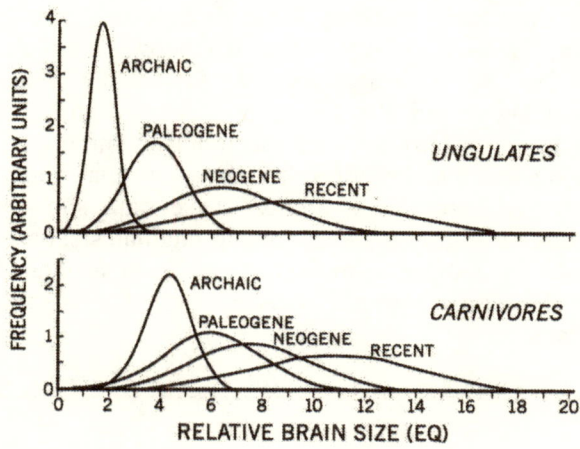

Figure 9.4: What Drives Progress in Evolution?: The ratio of brain size to body size in predators and their prey shows an adaptive "Arms Race." At any point in time, predators have relatively larger brains. Their prey must increase their brain size or face extinction. Then predators must increase their brain size, escalating the cerebral arms race into a literally vicious cycle.

Jerison provides further evidence for the arms race effect by examining the fauna of South America at a time when it was an island cut off from the rest of the world. Before more modern predators arrived via the Panama land bridge, the arms race between a separate line of predators (carnivorous marsupials) and their prey produced an identical pattern of progressive evolution as evidenced by increasing brain-to-body ratios (shaded area in Figure 9.5).

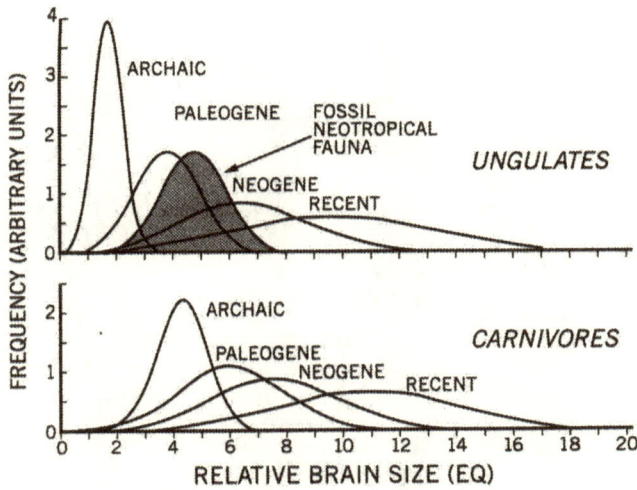

Figure 9.5: Independent Evidence of the Cerebral Arms Race. Before modern predators arrived in South America, when it was an isolated island, the now extinct carnivorous marsupials went through a similar predator-prey arms race in relative brain size (see shaded area).

Gould plays the contingency card for two reasons. The first is his desire to have evolution act in sudden jumps, consistent with his theory of punctuated equilibrium, rather than in gradual, ever more adaptive steps, in the Darwin-to-Dawkins British tradition—revolution rather than evolution. The second is that he wants to dispel forever the notion of evolution as a great chain of being—a long, determined process of steady, stepwise progress, from simple to complex, as depicted in Figure 9.6 from E. A. Hooton's classic *Up From the Ape*. As Gould explained, "Progress is a noxious, culturally embedded, untestable, nonoperational idea that must be replaced if we wish to understand the patterns of history.... Progress is not intrinsically and logically noxious. It's noxious in the context of Western cultural traditions." (Lewin, pp. 139 - 140)

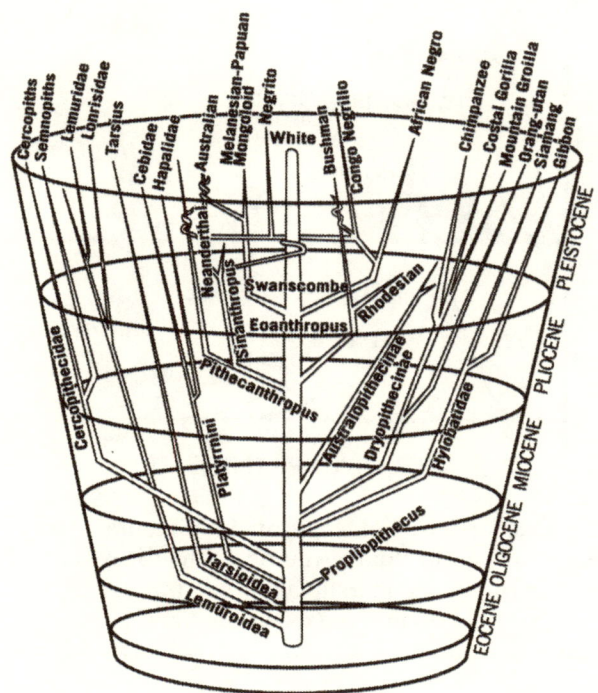

Figure 9.6: Invoking or Rejecting Progress: Human Evolution as shwon in anthropologist E. A. Hooton's classic, *Up From the Ape*. Figures such as this, with Whites at the pinnacle of the evolution, may account for S. J. Gould's reaction that "Progress is a noxious, culturally embedded, untestable, nonoperational idea that must be replaced if we wish to understand the patterns of history…."

Arguing that we would not see identical worlds is attacking a straw man – in this case, one that is dead and white. No one today, save creationists in their lampoons, thinks of evolution as some sort of cosmic time tunnel where a bunch of amino acids ooze into one end and Madison Grant, author of *The Conquest of a Continent* and *The Passing of the Great Race*, marches triumphantly out the other. The real questions we should ask are: "Just how different would these worlds be?" and "What are the key morphologies and behaviors that are selected for or against?"

When one turns from the evolutionary record to the record of human history, the use of contingency, and what in that discipline are called counterfactuals, lends itself to even greater abuse.

COUNTERFACTUALS: FROM "WHAT IF ..." TO "IF ONLY..."

I couldn't find counterfactual in the dictionary. This jargon term is a trifecta of Coulda, Woulda, and Shoulda, best captured by George Herbert's ditty:

> For want of a nail the shoe is lost,
> For want of the shoe the horse is lost,
> For want of the horse the rider is lost,
> For want of the rider the battle is lost,
> For want of the battle the war is lost,
> All for want of a horseshoe nail.

While contingency is offered as an alternative to determinism, the small and apparently insignificant changes leading to a cascade of wildly different scenarios, counterfactuals can act as determinism with a vengeance. As Niall Ferguson, one of the better practitioners of counterfactualism, points out in his introduction to a collection he edited called *Virtual History*, the "What if ..." of counterfactuals all too easily becomes the special pleading "If only...." "What if JFK hadn't been killed?" all too easily becomes "If only JFK hadn't been killed, we wouldn't have gotten into that terrible Vietnam war." Or, "What if FDR knew about Pearl Harbor?" becomes "If only the American voters knew what FDR was hiding from them, we wouldn't have gotten into the Second World War." Oliver Stone's "historical films" (*JFK, Nixon*) and Pat Buchanan's recent book on United States foreign policy (*A Nation, Not an Empire*) are less about what either of them thinks our policy should have been than about what they want our policy to be now and in the future. "Such veiled counterfactualism has been a striking feature of a great many recent 'revisionist' works of history." (Ferguson, 1997, p. 19).

The British Museum has an entire room devoted to counterfactuals, which it politely terms "Imagined Histories." The Smithsonian calls them "Alternative Histories." Gore Vidal has made a career of writing mixes of history (the way he thinks it really happened) and novels. His latest effort in the genre, *The Smithsonian*, takes place in that very institution. If that's not enough, you can also go to the *Uchronia: Alternate History web page* (www.uchronia.net) where you'll find an annotated bibliography of novels, stories, essays and other material involving the "what ifs" of history, known in the trade as alternate histories, allohistories, uchronia, counterfeit worlds, counterfactuals, and negative histories.

Counterfactual works abound. Ferguson's (1997) edited volume, *Virtual History*, contains chapters on England Without Cromwell, the world without the American Revolution, Britain 'Standing Aside' in W.W.I, Hitler's invasion of England, Hitler's defeat of the Soviet Union, the world without the Cold War, Camelot lives (i.e., JFK doesn't die), and the world today if the Soviet Union had survived. *What If?* (Cowley, 1999) contains counterfactuals by historians William McNeil (on how a plague saved the 10 tribes of Israel from being carried off by the Assyrians in 701 BC), John Keegan (on how Hitler could have won the war), and Stephen Ambrose (on what if D-Day failed), among others. *The Hitler Options: Alternate Decisions of World War II* presents 10 different scenarios by which *der Führer* (a perennial counterfactual favorite) could have pulled it off (odd that he didn't employ any of them).

Robert Sobel's (1973, 1997) *For Want of a Nail: If Burgoyne Had Won at Saratoga* and Niall Ferguson's (1999) *The Pity of War: Explaining World War I* fall into a different class of counterfactual. Sobel's *For Want of a Nail* traces the history of British America from 1763 to 1973. His 440-page tome includes fictitious maps, election results, economic trends, and a 16-page bibliography. A critical afterword, appearing in the (fictitious) North American Historical Association, shows that the book was meant as a thinking exercise to illustrate two major trends in the history and politics of Anglo-America:

> It is the author's contention that the peoples of British North America were fairly united prior to 1763. Although he does not say as much and in so many words, he implies that the North Americans of that day had two political traditions. The first of these came from Britain, and consisted of an evolutionary road to a more perfect commonwealth. The second, a child of the Enlightenment, was a utopian view of man and nature, to be found, in differing ways, in such thinkers as Voltaire, Rousseau, Montesquieu, and others of that period. Following the Seven Years' War, some Americans … held that the nation's future lay with the Empire, with Europe, with tradition, and with the evolutionary development of institutions. A second group, including the Adamses and Jefferson, were utopians, who believed in free will, held that man could be the master of his fate, rule himself, and wash away the abuses of centuries in a generation. These two groups and ideologies clashed in the Rebellion, which saw the victory of the "evolutionists," if we may call them that. The hero of this book is John Burgoyne, first Duke of Albany, who wins the Rebellion and then sets the new C.N.A. [Confederation of North America] along the evolutionary pathway. (p. 405).

Evolution v. Revolution. Sound familiar?

TOO MUCH OR TOO LITTLE?

Ferguson's *The Pity of War* is not really an alternative history. It is an all too real account of World War I, not in terms of a military or a diplomatic narrative, but rather in terms of major topics and a discussion of how much effect they had. Chapters examine specific questions about why W.W.I began, why it was fought the way it was, and why it ended as it did. The book is most illuminating when Ferguson gets down to counting literal "bang for the buck"—how much it cost each side to kill one of the other side's soldiers.

The Pity of War only becomes counterfactual in the final chapter, "Alternatives to Armageddon," where Ferguson challenges the reader to consider what would have been the result if Britain had "stood aside" (the British terminology at that time for not having entered the war against Germany). Ferguson's conclusion is that not only Britain, but Europe and the U.S., would have been a lot better off if the Brits had kept out. The war in Europe would have been a stalemate. No collapse of the Tsarist regime means no Bolshevik revolution and all that came with it. No humiliating defeat for the Kaiser's Germany means no Treaty of Versailles, no Hitler, no W.W.II, no Holocaust, no Cold War.

Again, Ferguson's book is meant more as a thinking exercise on policy options than an alternative history. To that end, he appeared on C-SPAN opposite Harvard professor Charles Maier and Daniel Goldhagen, author of *Hitler's Willing Executioners*. Goldhagen is so wedded to his thesis that the Holocaust was inherent in the anti-Semitism of the German people he went so far as to argue that even if there hadn't been a Versailles Treaty Hitler would have come to power, in Austria-Hungary no less, because of the world depression of the 1930s. Now there's a counterfactual for you!

Both Shermer and Pigliucci have used the famous example of Lee's Order #191 falling into Union general McClellan's hands to illustrate a counterfactual. I'll use one less well known—the service above and beyond the call of duty supplied by an American nurse before the Battle of the Bulge in World War II—to illustrate how I think the concept is limited and how it needs to be refined and improved.

In December of 1944 the Western allies were convinced it was only a matter of time before they crossed the Rhine and conquered Germany. Der Führer, ever the gambler willing to "think outside the box" (maybe that's why he's such a favorite with counterfactualists), gathered together whatever forces he could spare from the Eastern front for one last chance to win der Krieg. He attacked where the Allied forces had left themselves vulnerable, pushing forward to form a thinly defended "bulge." His army and elite Waffen SS troops, surprised the Allies by advancing through the lightly defended Ardennes forest. Hitler hoped to split the British and American forces, driving all the way to the port city of Antwerp. This would deprive each army of its supply base and produce a "Second Dunkirk" that he hoped would cause Britain to sue for peace. He could then turn his attention to the Eastern front.

The offensive was initially successful, especially against those sectors defended by relatively inexperienced American troops, and because bad weather kept Allied air forces grounded. Eventually it was halted far short of Antwerp. The Americans refused to surrender the vital crossroads of Bastogne, with which two names are forever linked. The first is that of the defending commander, General McAuliffe, who replied "Nuts" to the German request for surrender. (He may actually have uttered another four-letter word, but these accounts were written in the 1940s when "trash talk" was not in vogue.) The second hero is General Patton, whose 3rd Army eventually lifted the siege of Bastogne. (Who can forget actor George C. Scott in his Oscar-winning role ordering the military chaplain to provide a weather prayer so that he would have air cover for his advance?) But British military historian Sir Basil Liddell Hart discovered a third hero, actually a heroine, on whom I shall build my counterfactuals. When he interrogated the Wehrmacht commanders after the war one of them told him he could have taken Bastogne early on, but he had "dallied with a young American nurse, 'blonde and beautiful,' who held him spellbound." (Liddell Hart, 1970, p. 651, n. 2) Surely this unsung heroine was worthy of, if not the Purple Heart, at least some appropriate honor. In any case, let's consider what would have happened if the Panzer commander hadn't wasted vital time playing doctor.

<u>Counterfactual #1</u>. The plan goes exactly as Hitler hoped. The British troops are cut off in a second Dunkirk and Otto Skorzeny's commando troops produce widespread panic among the inexperienced American troops. The Churchill government falls and King George VI abdicates. A successor government with the Duke of Windsor restored as Edward VIII releases Rudolf Hess from prison and sues for peace. The shock causes FDR to die of a heart attack. He is succeeded by Vice President Henry A. Wallace, who was elected with him in 1940 but was denied renomination at the 1944 Democratic Party convention. The military ignores all orders from the hapless pro-Soviet Wallace and waits until the anti-Soviet Harry Truman assumes the presidency in January of 1945. Listening to his generals and admirals, Truman turns America's war effort to defeating Japan and blocking any Soviet drive in the Pacific. Now fighting on only one front, Hitler outmaneuvers the Soviet army as he did in 1941 and wins back the Eastern front. As in Robert Harris' novel *Fatherland*, an aging Hitler eventually shakes hands with American president Kennedy—the also aged

Joseph Sr., the non-interventionist. The big debate among future historians is when, where, and how the European Jews disappeared.

But even supposing that the officer-nurse tryst didn't take place, would all (or any) of these events follow inexorably? It's just as likely that after his victory in the West, Hitler would have once again gambled in the East and this time lost big. With Britain and the U.S. out of the war, victorious Soviet armies would have eventually taken all of Germany, and the "Iron Curtain" (a term Josef Goebbels used before Winston Churchill by the way) would have been the Atlantic Ocean. Rather than Harris' *Fatherland*, the result would have been more like Orwell's *1984*.

Counterfactual #2, #3. Any number of other scenarios might be imagined. Fearful that the war would now drag on forever, a successful officers' plot might have killed Hitler and sued for peace with the West (thus putting the Iron Curtain on the Vistula, rather than the Elbe) or, if rebuffed, with the Soviet Union, or with Russian generals inspired to eliminate Stalin (thus putting it on the Rhine). Or it could have made very little difference at all in the grand scheme of things and just meant that more lives would be lost and cities ruined. Especially when we are dealing with human beings who make conscious decisions, often based on past experience, events do not flow inexorably.

Counterfactuals are useful as thinking exercises that help us understand how events are interconnected. They become more meaningful as they become more mathematical and therefore testable. Tetlock and Belkin (1996) and Schroeder (2000) describe methodologies for determining what constitutes reasonable and unreasonable counterfactuals. The best examples to date are the cliometric (measuring Clio—the muse of history) work of Robert Fogel. His analyses were at first disputed, largely because his new methodology challenged two sacred cows of historians and economists (particularly those on the American left): Fogel concluded that without the Civil War slavery would have continued to be economically profitable, and that railway construction was not a sine qua non for U.S. industrialization. (Fogel got the last laugh, receiving the Nobel for economics for his work in 1993.)

A QUINELLA—CHAOS AND COMPLEXITY, OR CONTINGENCY AND COUNTERFACTUALS

Chaos and complexity are important concepts in mathematics and physics, but we should look skeptically at their introduction into other fields, especially in books with titles that sound like a cross between a drive-in movie from the 1950s and a self-help seminar: *Complexity: Life at the Edge of Chaos*; *The Collapse of Chaos: Discovering Simplicity in a Complex World*; *Frontiers of Complexity: The Search for Order in a Chaotic World*.

Rather than an alternative to determinism, contingency can be a disingenuous form of determinism when it is applied in counterfactuals as some sort of 19th century railroad switch, diverting the train of history to one and only one "entirely different but equally sensible outcome." A better metaphor is to think of events, in either evolutionary or historical time, as cars swerving between the fast and exit lanes during rush hour traffic. Odds are the total traffic flow will be affected in all the lanes, with lots of fender-benders and crashes as the lanes accordion in and out. Drivers in many, but not all lanes will make adjustments. Most will make it home safe, if delayed. Some will come home bandaged. And some will end up toe-tagged and body-bagged. Which drivers make it and which ones don't will depend on their individual driving ability, the performance of their car, and luck as well. But tomorrow there will be another rush hour (Figure 9.7).

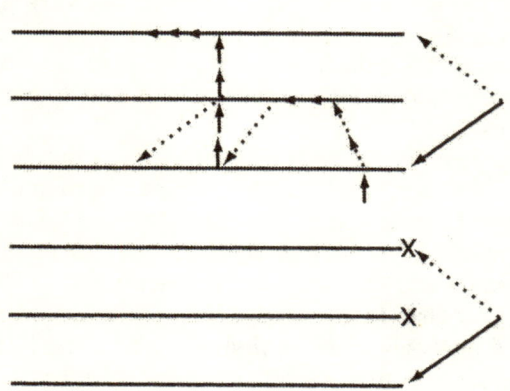

Figure 9.7: Rigid Determinism versus Alternative Outcomes. Rather than alternatives to determinism, contingency and counterfactuals are disingenuous determinism when viewed as a 19th century railroad switch that inexorably diverts the course of evolutionary or human history on some other course (top). They are useful as thinking exercises where multiple options can be considered and mathematically evaluated (bottom).

Michael Shermer likes to point to the huge file of "Theories of Everything" articles submitted to *Skeptic* to illustrate how theories that attempt to explain everything usually explain nothing. Outside of the physical sciences and mathematics, chaos and complexity may be just analogies and metaphors, theories of everything that are little more specific than the daily horoscope or the messages in fortune cookies. But uttering the words "chaos" and "complexity," especially in the right tone of voice and at the right time, can get your way paid to an academic conference in some fashionable part of the world (where the most important conversation will go on over drinks), or land you that management job (that requires only meeting the minimal qualifications). And you can make book on that!

References

Bak, P. 1996. *How Nature Works: The Science of Self-Organized Complexity.* New York: Springer.

Cohen, J. and Stewart, I. 1994. *The Collapse of Chaos: Discovering Simplicity in a Complex World.* New York: Viking.

Conveney, P. and Highfield, R. 1995. *The Frontiers of Complexity: The Search for Order in a Chaotic World.* New York: Fawcett.

Cowley, R. (ed.) 1999. *What If? The World's Foremost Military Historians Imagine What Might Have Been.* New York: Putnam.

Ferguson, N. 1997. *Virtual History: Alternatives and Counterfactuals.* New York: Basic Books.

Ferguson, N. 1999. *The Pity of War: Explaining World War I.* New York: Basic Books.

Jerison, H. 1972. *The Evolution of the Brain and Intelligence.* New York: Academic Press.

Lewin, R. 1992. *Complexity: Life at the Edge of Chaos.* New York, NY: Macmillan. 74.

Liddell Hart, B. H. 1970. *History of the Second World War.* New York: Putnam, 651, n.2.

Gould, S. J. 1989. *Wonderful Life: The Burgess Shale and the Nature of History.* New York: Norton.

Russell, D. 1992. *An Odyssey In Time.* Toronto, Canada: University of Toronto Press.

Schroeder, P. 2000. "Embedded Counterfactuals and World War I As An Unavoidable War." Paper presented at the 2000 Annual Meeting of the American Political Science Association to appear in Tetlock, P., Lebow, R., and Parker, (eds.) *Unmaking the West: Counterfactual Thought Experiments in History.*

Sobel, R. 1973/1997. *For Want of A Nail: If Burgoyne Had Won at Saratoga.* London: Greenhill Books.

Tetlock, P. and Belkin, A. 1996 *Counterfactual Thought Experiments in World Politics.* Princeton, NJ: Princeton University Press.

Part III

The Race-IQ-Genetics Debate

Frank Miele

CHAPTER 10

FOR WHOM
THE BELL CURVE TOLLS

CHARLES MURRAY
ON IQ, RACE, CLASS

AN INTERVIEW WITH THE AUTHOR OF
THE BELL CURVE

Charles Murray achieved the impossible, or at least the highly improbable. He has co-authored an 845-page book, filled with figures, tables, references, and appendices loaded with multiple regression analyses, that is also the most controversial book in America. It has been panned by many outside the intelligence testing community and by some within. Commentators from the left, right, and middle have taken their best shots, and leaders of both major political parties have rung in with scathing attacks, even while admitting they had not read the book. Despite the brouhaha, or perhaps because of it, *The Bell Curve* made it to the *New York Times* top-10 nonfiction best-seller list. As Ted Koppel put it on *Nightline,* *The Bell Curve* will be like Clinton's health plan: no one will actually read it but everyone will form an opinion of it.

Charles Murray is no stranger to controversy. His previous book, *Losing Ground: American Social Policy, 1950-1980,* argued that the Great Society programs of the 1960s not only did not help the poor, they often made things worse, and suggested that the welfare system be abolished. The *New York Times* called *Losing Ground,* "The [Reagan] Administration's new 'bible.'" *Losing Ground* was but a prelude.

In *The Bell Curve: Intelligence and Class Structure in American Life* (co-authored with the late Richard J. Herrnstein, author of the earlier controversial *IQ in the Meritocracy,* and co-author with James Q. Wilson of *Crime and Human Nature*), Murray has pushed the envelope of public political discourse to its breaking point. He has now been dubbed by the *New York Times Magazine* "America's Most Dangerous Conservative." When editor Andrew Sullivan ran an excerpt from *The Bell Curve* in *The New Republic,* its entire editorial board rose in revolt. But a group of leading researchers in the field of human intelligence published a statement in the *Wall Street Journal* agreeing with the factual basis of *The Bell Curve*.

Frank Miele

Herrnstein and Murray argue that IQ is real; that it matters (ever so much more as society becomes more egalitarian and technological); that it is somewhere between 40% and 80% heritable; and that it relates to not only school performance, but to jobs, income, crime, and illegitimacy; and that it cannot be ignored in any meaningful look at America's future. But the most explosive of *The Bell Curve's* arguments is that some of the difference in mean IQ scores between the white European population of the United States and the African-American population (one full standard deviation of 15 points) is probably attributable to genetic factors. No one in the field disputes this difference. The argument is over *why* the difference exists and, of course, whether and how it can be reduced. (Read the now-infamous Chapter 13 of *The Bell Curve* for yourself. It is a lot more tentative and nuanced than popular denunciations of the book may have led you to believe).

Charles Murray is a graduate of Harvard with a Ph.D. in Political Science from MIT, and currently is a Bradley Fellow at the American Enterprise Institute in Washington, DC. The late Richard J. Herrnstein received a Ph.D. in psychology from Harvard, where he taught from 1958 until his death last autumn. He held the Edgar Pierce Chair in Psychology (the oldest such chair in America).

Skeptic interviewed Charles Murray and found him to be as calm in explaining his positions as some of his critics have been apoplectic. If there was any trace of anger in Murray, it was not for the underclass but for former colleagues who, as he put it, "ran for the high hills," and for the Cognitive Elite, who he feels have undermined the country that provided them the chance to rise to the top in the first place.

This interview should be compared and contrasted with that of Professor Robert Sternberg, author of *The Triarchic Mind* and the editor of *The Encyclopedia of Intelligence* in the next chapter.

Here is what then "America's Most Dangerous Conservative" had to say when I interviewed for *Skeptic* at the height of *"The Bell Curve Wars."*

184

Figure 10.0: Charles Murray

Miele: In your book you present a summary of the current evidence on IQ on pages 22 and 23. Snyderman and Rothman's 1991 book, *The IQ Controversy,* surveyed experts in the field, and just yesterday *The Wall Street Journal* contained a 25-point statement by experts on intelligence. Based on those it seems your summary represents the consensus of experts in this field, even on the controversial issue of the involvement of genetics in the black-white difference in intelligence. As skeptics, we are skeptics of everything, including psychology. If we get this great a controversy over what looks like consensus, is psychology really a science in the same sense as physics?

Murray: I'm not comfortable with a blanket statement saying yes or no. But I think we can talk specifically about the basis for those statements in *The Wall Street Journal* and the book, which is certainly based on the kinds of methods that fall under the scientific method – falsifiable hypotheses, the use of predictions, etc. A test is valid insofar as it predicts something of interest – a criterion measure to use the jargon of the trade. More than most of the other social scientists, psychologists and psychometricians are prepared to have their results tested against classical statistical criteria of validity, reliability, and reproducibility.

Miele: One of the arguments would be that most of the analyses you and psychometricians have done is correlation, as opposed to causal analysis that we do in physics. Does that mitigate the strength of the scientific conclusions?

Murray: We do not have accessible to us the same kind of control over our phenomena that physicists often have. However, having said that, there remains a black box where the cause hides that we cannot open up and look at. But one can eliminate a number of alternative hypotheses and transform correlational statements into ones which certainly have some causal persuasiveness. Example: the use of regression analysis, which is the all-purpose tool of the behavioral and social sciences these days. Let's take an example from *The Bell Curve*. The dependent variable is whether a person is below the official poverty line. And you introduce as independent variables a variety of candidate causes, chief among these are socio-economic background, education, race, occupation, and then you throw IQ into that. If after looking at a variety of these other things which both theory and common sense say should have some bearing on whether a person ends up in poverty, but one ends up with a large, statistically independent role for IQ, it seems to me to make a causal statement that IQ looks like it's a cause of poverty, it is a reasonable thing to do.

By the way, when you actually read the book you will see that we typically word things in this cautious kind of way.

Miele: But could not someone say that in correlational analysis it is not really proper that you are not randomly assigning people to socio-economic status groups, or racial background, or whatever; are you and Herrnstein doing anything different than what is the common procedure used in regression analysis in, say, voting behavior, or anything of this sort?

Murray: The analyses we conduct in the book are garden-variety regression analyses. There is, however, also a body of work which does use randomly assigned experimental and control groups that reflects on a lot of the issues we talk about in the book, which then begins to look more like experiments done in the hard sciences.

Miele: Can you be more specific?

Murray: Arthur Jensen's work with regard to reaction time. This is a matter of eliminating a lot of alternatives and trying to understand what's going on with IQ scores. This is where you have an apparatus with six buttons, and you have your finger on a button, and when a light goes on one of the other buttons, you move to that button and push it. The reason why this kind of experiment is useful is it gives us insight into something that has no known relationship to IQ, insofar as you are simply asking someone to move his or her finger to push a button. But it turns out that this reaction time is not only correlated with IQ scores, it is correlated with the general intelligence factor, *g*.

The main point is this. You have now made some headway into trying to understand what is going on with this thing called an IQ score – does it have a physiological basis?, etc.

I'm not going to apologize for our use of regression analysis in the book. Everyone uses regression in the social sciences. If you want to say that social scientists are the astrologers of the 20th century and that they don't have the methods of science open to them and we thus can't take them seriously, fine. But unless you are prepared to make that argument, the science in *The Bell Curve* using regression analysis is very much in the middle of the mainstream.

Miele: Stephen Jay Gould, in his *New Yorker* review, gives a four-point summary of your argument about intelligence: (1) it is a single number; (2) it is capable of ranking people in a linear order; (3) it is genetically based; (4) it is effectively immutable. Gould goes on to say that if any of these premises is false, the entire argument of *The Bell Curve* collapses, and he concludes that: "The central premise is false and most of the foundations are." Now, how do you square what Gould, Gardner, and Sternberg have said in their reviews of *The Bell Curve* with your own summary of the book on pages 22-23, as well as with Snyderman and Rothman and *The Wall Street Journal* statement. One of you has got to be wrong.

Murray: Stephen Jay Gould is recycling the same argument that appeared in *The Mismeasure of Man* in 1981. It was a very influential book in terms of the lay population and lots of people out there have their opinions formed about intelligence by *The Mismeasure of Man*, which included two types of information, one considerably more useful than the other. The first type was a history of intelligence measurements in the 19th and early 20th centuries, in which people made mistakes. (I seem to recall that physicists used to believe in something called aether.) There were phrenologists and others that we can now look back at and poke fun of. Fine, the problem is that in the same way that physicists are not criticized now for something other physicists did in 1910, so also has psychometrics made some strides since 1890 and 1900 and 1910.

The second thing is that Gould tried to make the same arguments for modern psychometrics, a lot of which were based on trying to demonstrate that the general mental factor g is a statistical artifact. The contrast I want to draw is between the attention that Gould's book got in the media and what happened in the scientific literature. Basically, there were a few perfunctory and rather derisive mentions of his treatment of factor analysis, and work went on without a break.

To put it more specifically, factor analysis is subject to a variety of kinds of problems because you can make different assumptions about how to create the factors. One can even, for example, demand of the algorithm that it produce factors which do not load on a single dominant factor. The problem with this is, as Dick Herrnstein use to say, you can make g hide but you cannot make it go away.

Miele: Sounds rather like Joe Louis. Let me go back to Gould's four points. Is there any one of those that you think is not a fair and accurate statement of what you said?

Murray: All four of them.

Miele: So you are not saying intelligence is a single number?

Murray: No. In *The Bell Curve,* we say of the IQ score, first, there have been a variety of ways to try to come up with independent mental factors. That has been a failure. On the other hand, there have been a variety of ways in which there are distinctions among different types of intelligence that are useful, such as the distinction between verbal, visual and spatial intelligence. Further, we talk about the different ways these different skills lead to success in occupations. And we talk, somewhat sympathetically, about the notion that there are, in Howard Gardner's words, multiple intelligences. We are a little dubious about applying the word "intelligence" to them, but we are very sympathetic that there are large domains of human talent that are not encompassed in the word "intelligence."

Miele: One of the complaints about the Snyderman and Rothman survey, *The Wall Street Journal* statement, and your own survey of the literature, is that you are working in that standard psychometric paradigm, but that is yesterday's news. The real forefront is Sternberg's approach to practical intelligence and Gardner's seven intelligences. You are sticking with something that is a very small portion of the discussion, so naturally you are going to get consensus.

Murray: Let me make a couple of other points about intelligence. One, the general mental factor, *g,* is very robust. You can take all kinds of different ways of creating your factors, and you will always get *g*. It doesn't matter whether you rotate the matrix orthogonally, or obliquely, or whatever else, you always get the same thing. The second major point is that when you try with factor analysis to produce a situation where you do not have a general mental factor *g,* guess what? All the factors are correlated, which goes back to Spearman's initial insight – why are the different measures of mental ability so consistently correlated with each other? What's going on here? The answer is: there is an underlying general factor. That does not mean that it blocks out a whole territory of human talents or intelligence, and we say so in the book.

Gardner has made a variety of assertions about intelligence which, if true, are falsifiable. He is not only saying there are different talents, which Dick Herrnstein and I would agree with, he is also saying they are independent. With something like kinesthetic talent, which is quite physical, this is easy to say. It gets harder to say when he talks about interpersonal skills versus verbal skills. If you are going to make that kind of statement, the next logical step is to come up with measures of these different talents and demonstrate that they are, in fact, independent.

Miele: So you are saying that some of these disagreements are empirically testable?

Murray: Yes, and Gardner has consistently been unwilling to subject his own work to that kind of empirical defense. He has stood apart from quantitative attempts to describe what he is doing and to enable other researchers to replicate it. Of all the different types of intelligence that Gardner wants to treat as co-equals, there is only one kind that will put you in the retarded class – namely the plain old-fashioned general mental ability. If you are kinesthetically challenged, teachers and guidance counselors do not get real worried about you. If you are kinesthetically challenged you may be the last person chosen for the baseball team, but you can go out and make a success of yourself in any number of ways. If you are intellectually challenged, however, you have a general disability that is pervasive over all kinds of ways.

Miele: I read in a biography of Muhammad Ali that when he took his draft tests he had an IQ below 80. Now, if I make a mistake writing, my spell-checker will fix it for me. If you make a mistake, Stephen Jay Gould may beat you up in the press. If Muhammad Ali made a mistake, he was flat on his back. This man was making split-second decisions of the first magnitude.

Murray: If you are five standard deviations out on the right hand [positive] edge of the curve in kinesthetic ability, then certain possibilities open up to you. But if you are low in kinesthetic ability, it doesn't make much difference to you in life. If you are a Muhammad Ali and you possess extraordinary physical talent, you have other avenues that will open up to you. But this example illustrates another important point, which is that Muhammad Ali is not a blithering idiot. There is nothing in his public utterances, his charm, his presence, his charisma, and all the rest of that, that is inconsistent with a measured IQ in the high '70s.

Miele: One of the things that Gould takes you to task for is that you do not report the scatter on your regression lines, and that the *r squared* values do not appear in the body of the book, but in the appendix. Can you tell us what those terms mean and why he thinks they are so important? And what is the usual practice here – have you and Herrnstein done something different than what would be done in a book on, say, political voting behavior?

Murray: Correlation is denoted by *r,* and in ordinary regression analysis *r-squared* – the square of the correlation – reflects the proportion of the variation in the data that is explained by the set of independent variables that are in the regression analysis.

Two points. First, the book is exemplary in opening up the section in which we present these regression analyses by explicitly pointing out in the body of the book that the *r-squared* is small. That is in the very first pages of the whole presentation. It is also exemplary in a book aimed at a general audience that we specifically include an appendix with a print-out of every single analysis presented in a graph in the body of the book. This is something you will not find in most books aimed at a general audience, including *The Mismeasure of Man.* Dick and I presented far more statistical information than is ordinarily presented in a book such as this.

Second, I don't know how much Gould knows about regression analysis. When you are using logistic regression analysis, in which the dependent variable is a nominal variable, in our case a dichotomous yes-no, is the person below the poverty line, yes or no? Whenever you have a dichotomous dependent variable, *r-squared* becomes very difficult to interpret, particularly with rare phenomena. For example, if you have a situation, as in the case of poverty, where 87% of the population is not impoverished and you only permit two values in your dependent variable (i.e., above or below the poverty line), you are going to get a lot of noise in the data, which means *r-squared* becomes very difficult to interpret for technical reasons.

Miele: One of the criticisms, then, would be that the IQ isn't that effective. There is a lot of noise, so why are you saying it is so important?

Murray: Let's use poverty as an example. For poverty the *r-squared* is .10. So we can explain only 10% of the poverty, so obviously IQ cannot be very important, can it? Well, if you then go back and take a look at the probability that a person is going to be in poverty if that person has a low IQ, you find out that among whites, the probability of being in poverty if you are in the bottom 5% of IQ is 30%. The probability of being in poverty if you are in the top 5% of IQ is 2%. Furthermore, when you take into account education and socio-economic status, the magnitude of the difference in probability of being in poverty is not much reduced. How can this be if you can only explain 10% of the variance? It goes back to the ways in which logistic regression equations in which *r-squared* is not nearly as interesting as the magnitude of the effect that IQ has on the probability of being in poverty. It's like a slot machine where a small percentage difference can add up when the numbers are big enough. And this applies across the range of the analyses we report.

Miele: Let's try to cut this another way. If you get so much predictive value from using intelligence, just an IQ score, how much do you add by getting these other measures of socio-economic status or whatever. I.e., what's the sequence? What's the biggest predictor, and how much do you add by cranking the others into the equation?

Murray: It depends on the phenomenon you are looking at. Once you introduce IQ, the importance of socio-economic background is much reduced. Often the role of socio-economic background disappears altogether when IQ is in the equation. Conversely, introducing socio-economic background into the equation often reduces the independent role of IQ by only a very small amount.

Miele: Let's talk about the cognitive stratification that you discuss in Part I of your book. Secretary of Labor Robert Reich gave a speech on the "Anxious Class," where he said:

> Contrary to some theorists, our destinies do not reside in our genes. Study after study shows that skills can be learned. Every year of education or job training beyond high school, whenever it occurs in life, increases average future earnings by 6% to 12%. GNP is not simply a matter of DNA. Most Americans are on a downward slide not because of genetic deficiencies but because they lack the learnable skills to prosper in an economy convulsant with change.

But the picture Reich paints in his book, *The Work of Nations,* is actually very similar to your own: a high-end cognitive class that is doing great, a bunch of worried people in the middle saying "where's that job my father had?;" and an underclass at the bottom that is falling downward in free-fall.

Murray: Yes, and we make a reference to Reich's work and we point out that he is more optimistic about the role of education than we are, but there are great similarities. There's also a lot of similarity between our vision of what's going on and Mickey Kaus' vision in his book *The End of Equality.*

Miele: Apparently Mr. Kaus did not see that similarity or chose not to see it when he wrote a scathing attack on *The Bell Curve* in the *New Republic.*

Murray: Mickey Kaus ran for the high hills.

Miele: And on this subject of what society can do to bring up this underclass, you have been something of a godfather of the get-rid-of-welfare movement before you got into talking about IQ. Conservatives like Newt Gingrich, Buchanan, Kemp, and Bennett have also talked about getting rid of welfare, but they have rejected *The Bell Curve's* analysis of IQ. Does one of these follow from the other?

Murray: You know, in all 845 pages of *The Bell Curve,* we talk about getting rid of welfare in one sentence. We have a single sentence in Chapter 22 in which we talk about the ways government should get out of the business of encouraging any group of women to have babies, whether they be smart or dumb. And we generally urge that policies that subsidize birth be ended. It is one sentence. A single sentence. And one does not follow from the other.

Miele: Even if there were no heritability to intelligence, no racial difference in any of these things, you would still be in favor of getting rid of welfare?

Murray: Yes, absolutely.

Miele: One question we might ask about your book is: why this book now, and why the controversy that surrounds it? Is this a case of a bad economy, an anxious public, and so we are blaming the victims and scapegoating those least capable of defending themselves? Aren't you giving aid and comfort to a revival of racism?

Murray: Why is it published now? A better question is: why was it not published in the last 30 years? There has been a collective intellectual cowardice about understanding the role of intelligence in understanding social problems. Let's take one example. Child neglect is one of the most rapidly growing social problems we have. How many thousands of people make their living either writing about problems of child neglect and abuse, and so are advocating for new laws, etc.? Well, as every parent knows without reading anything about IQ, there is a plausible relationship between intelligence and child abuse. Which is to say, any parent knows if a child has had a fever for 24 hours and hasn't been taking in liquids, you make a calculation that this has gone on too long and we've got to get this kid to a doctor, etc. Any parent knows that child-proofing a home takes foresight and thoughtfulness – it takes a certain amount of IQ. With that plausible relationship in mind, the failure of social science and politicians alike to confront the possibility that low IQ is an important risk factor in child neglect is scandalous. Every single bit of evidence that does bear on this says that IQ is a great big factor in child neglect.

Miele: Couldn't someone take your arguments and say "we need *more* redistribution programs, not less, because these people cannot help themselves"? Haven't you knocked the bottom out from the conservative pull-yourself-up-by-your-own-bootstraps ideology?

Murray: You put your finger on something that Dick Herrnstein and I also thought about from the time we began working on the book. It is something that my friends on the right were concerned about. They said, "Look, this type of material lends itself to all sorts of reasons to have a more interventionist state, and more welfare, and more redistribution, not less." We knew that. That is one of the major reasons for saying that it doesn't really hang together that this book was designed to foster a political agenda. It can be used by both sides.

The other point is that you have just described is why Dick and I open up Chapter 22 with seven or eight pages of political philosophy. We say to the reader very explicitly "What we have just described for you could play out in any number of ways politically. Therefore what we are going to do is describe to you our own political predispositions, which have nothing to do with IQ, why we hold them, and having told you those predispositions, then we will tell you our strategic view of what ought to be done."

Frank, I challenge readers of *Skeptic* magazine to go to any other book with policy recommendations by liberals which contain such an explicit, open, candid description of the author's political bias.

Miele: I'd like to go further out on that limb. One might argue that *The Bell Curve* challenges the whole tradition that many people identify as American – namely equality. Do you find that your conclusions better fit a different vision of the universe that sees humanity as continuous with the rest of existence rather than as created in the image of God, and the Goddess Fortuna working her wiles through DNA?

Murray: Our vision is Jeffersonian. Up until 30 years ago, in the early 1960s, Dick and I would have been describing a vision of America that everyone would have said, "of course." It is a vision in which we say that people bring different things to the table. The important thing is that everyone be given the opportunity to go as far as his or her temperament, energy, characteristics, and intelligence will take them. The crucial factor in coming up with a harmonious society is not equal outcomes, it is abundance of opportunity. We are talking straight out of a tradition that until 30 years ago had virtually no intellectual challenge. It is only in the last 30 years that people have lost sight of those fundamental tenets of the American idea. And Thomas Jefferson would read *The Bell Curve* and, I like to think, nod his head in approval. He believed there was a natural aristocracy that would make the republican experiment work. Personally I don't like the term "natural aristocracy" because I don't think the cognitive elite that we have now is all that great.

Miele: Along these political lines, with your previous work in many circles you have been the intellectual darling of the conservative anti-welfare crowd. But now that your book has stirred things up, do you feel that your former allies and friends are now running away from you, and how do you feel about that behavior?

Murray: I assumed when the book came out that a lot of people that used to think I was really neat would now say, "Charles Who?"

Miele: Has that happened?

Murray: I'm surprised at the extent that it has not. I thought that my political life would end. There seems to be a reflexive, almost deep inner panic in an awful lot of people to be on the right side of *The Bell Curve* issue. And the right side is being perceived publicly that you are *shocked* that these authors would suggest that intelligence has an important role in social problems, *shocked* to think that anyone would still suggest there are differences among the races in intelligence. (See Figure 10.1) I've seen people who I thought were both smart and honest, lie when it comes to the book.

Miele: Can you still be friends with these people?

Murray: No.

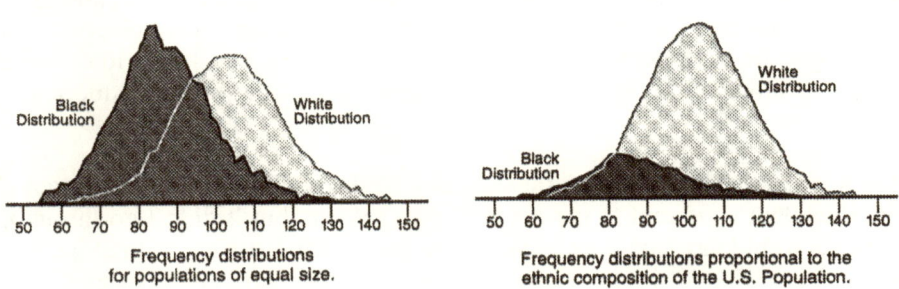

Figure 10.1: The "Flash Point" of The Bell Curve wars — The Distribution of IQ in the Black and the White populations of the United States.

Miele: As I've said, at *Skeptic* we are skeptical of everything. Given your experience do you think that the American political process can deal with the fact that *Homo sapiens* is a biological species, subject to the same laws of evolution and genetics as other animals? Can a democracy deal with the information in *The Bell Curve?*

Murray: Actually, I'm optimistic on this score. This book has created in the news media a type of hysteria, where it has been denounced not just as wrong, but as evil and misguided. But there are now over 400,000 copies of the book in print, and as my wife points out, correctly I think, people do not plunk down $30.00 to buy a pseudoscience, racist tract. They just don't do that. They are reading the book and talking about it.

I think what has happened to American intellectual life is that we have undergone a temporary aberration – 30 years, short as these things go –

whereby we have tried to deny all sorts of realities about human biological characteristics. The best thing about this book is that these issues have been taken away from the chattering classes. They are now out there in public discourse in a way that is going to provide cover for a lot of good scholars who want to talk more openly about these issues but have been reluctant to. I'm Panglossian in my optimism.

Miele: What happened the last 30 years?

Murray: What happened in the 1960s, and now I'm citing from *Losing Ground,* was a fundamental change in the view of how society works and what individual responsibility is, and this includes everything from education to law enforcement to the use of lawsuits, etc. It was a very widespread, but I think temporary, change that we are just now beginning to recover from and I think one of the lessons of this most recent election has nothing to do with people wanting a middle-class tax cut. It has to do with people wanting to return to a much more original view of how America is suppose to work.

Miele: Let me follow up on that. Hillary Rodham Clinton was in charge of the President's attempt to get a health care reform, but it didn't go through. No one would say that was because she didn't have sufficient intelligence, energy, knowledge, whatever. When I see your idealistic vision of what you would like to have in America it doesn't seem realistic. You are being Panglossian.

Murray: I was Panglossian about these issues getting into the public dialogue. Now let me shift to being the pessimistic curmudgeon, which I'm much more comfortable being! And that has to do with looking ahead to the long term. Hillary Rodham Clinton is the personification of what worried Dick Herrnstein and me about the cognitive elite. I'm sure she has a high IQ score. She, and for that matter her husband Bill, are both examples of people who by the age of 18 had been siphoned off into elite colleges and have spent the rest of their lives interacting with other people very much like them – the cognitive elite. And what happened with Clinton's health care program is a classic example of what happens when the cognitive elite has been talking to itself too long, and thinks it knows what's best for everyone. In this case, fortunately, they were derailed.

In the longer term what scares me is that the cognitive elite is, indeed, a powerful enough force to continue to rig the rules of the game. We are in favor of deregulation and decentralization, but I'm afraid the cognitive elite are going to make these things very difficult to carry out.

Miele: This bewilders me. You seem to say two different and possibly contradictory things. One, *The Bell Curve* finds the tremendous advantages that high IQ people have and which can be interpreted in a very elitist manner. Then, the libertarian Charles Murray emerges and says, "Well, the average Joe can run his life better than anyone else." How can you have it both ways?

Murray: Because running one's life is a matter of making all sorts of choices, and the satisfaction one achieves from running one's own life are inextricably linked with having been the person to make those choices. Someone with an income of $30,000 a year who made it himself, I submit to you is a happier man than someone who got that same $30,000 unearned, whether it comes from welfare or trust funds. People running their own lives, taking responsibility for their own actions, that's the way human beings are wired to live satisfying lives.

Miele: You are not a determinist. You are not saying everything is in the genes. You think free will is a meaningful concept.

Murray: Yes, and so did Dick Herrnstein, who was a student of B.F. Skinner.

Miele: Who didn't!

Murray: Yes, and Dick evolved a lot from his days as a behaviorist. One of the most difficult things to get across to people is that one may talk about genes playing an important role without being forced into anything resembling a determinist view of the world. But it is a contradiction only in this sense. The people who run their own lives are not necessarily going to make decisions that maximize anything in terms of some external source of comparison. In the Hillary Rodham Clinton world they might look at the things you have done or the choice you have made, and say, "No, no, if you had done this other thing you would have had more money, you would have had more security, etc." I'm saying that a lot of the basis for deciding whether the decisions one makes in running one's life are right or wrong has nothing to do with these types of external criteria.

You've asked very difficult questions that are hard to answer in a few sentences, but they are good questions.

Miele: Thank you. Is there anything you would like to add in conclusion?

Murray: I've enjoyed the interview. The only thing I would add is my own unhappiness at the way that Dick Herrnstein's name has been eclipsed. As

I've said to Susan Herrnstein, she would not be pleased to have Dick being called all the names I have been called over this issue. I have confidence that in five years from now, and thereafter, this book will be seen as a major collaboration between a political scientist and a psychologist. Dick loved data and respected the scholarly ideal of getting it right, absolutely right. When working with Richard J. Herrnstein the principle of *Veritas* [the Harvard motto] is applied in excruciating detail.

Miele: Thank you very much for taking the time to speak with us.

REFERENCES

Gardner, Howard. 1993. *Multiple Intelligences*. New York: Basic Books.

Gould, Stephen Jay. 1981. *The Mismeasure of Man.* New York: Norton.

Herrnstein, Richard J. and Murray Charles. 1994. *The Bell Curve*, New York: Free Press.

Kaus, Mickey. 1992. *The End of Equality*. New York: Basic Books.

Murray Charles. 1984. *Losing Ground.* New York: Basic Books.

Reich, Robert. 1991. *The Work of Nations*. New York: Knopf.

Snyderman, Mark and Rothman, Stanley. 1990. *The IQ Controversy*. New Brunswick, NJ: Transaction.

Sternberg, Robert. *The Triarchic Mind.* 1988. New York: Viking.

The Wall Street Journal. December 13, 1994. "Mainstream Science on Intelligence." P. A-17.

GLOSSARY

Bell Curve—in a statistically **Normal Distribution** (aka the "Bell Curve"), almost all the individuals fall within three **Standard Deviations** of the mean. On most IQ tests, the mean is 100 and the standard deviation is 15. About 34% of the individuals fall between the mean and +1 Standard Deviation; another 34% between the mean and -1 Standard Deviation. On the Wechsler Adult Intelligence Scale (WAIS) the mean IQ is 100 so 68% of the population has a WAIS IQ between 85 and 115. For the entire population, 99.7% have IQs between 55 and 145. IQs below 55 and above 145, therefore, are extremely rare.

For whatever reason(s), the mean IQ of African-Americans is about 1 Standard Deviation below that of the mean of the European-American population (that is, IQ 85 v. IQ 100). However, the full range of IQ scores is found in both groups, and it must be kept in mind that the bell curve represents a population. Individuals cannot be judged by their group. Also, the average difference between siblings born of the same parents and raised in the same home is about 12 IQ points.

Correlation—A statistic that ranges from -1.00 to 0.00 to +1.00, that measures how closely one variable (e.g., an IQ test score) is related to some other variable (e.g., income, or being above or below the poverty line). A correlation of 0.00 means that one variable predicts the other variable no better than chance; +1.00 means that it predicts the second variable perfectly; -1.00 means that the first variable predicts exactly opposite to the order for the second variable (that is, the higher the score on one, the lower the score on the other). Correlations are never 0.00. +1.00, or -1.00, but a fraction somewhere in between, represented by *r*, such as *r* = .34 or *r* = -.56.

When the correlation is squared (termed *r-squared*), it tells us the percentage of the observed variation in one variable that can be explained statistically in terms of the other. The remaining variation, that is, 1 minus *r-squared*, is the result of other variables, chance, or error. For example, suppose the correlation between alcohol consumption (per body weight) and reaction speed is .50, then 25% of the variation in reaction speed can be explained in terms of alcohol consumption and the other 75% is attributable to other factors.

Factor Analysis—A mathematical method used in psychometrics that reduces a matrix (or table) of the correlations between a number of variables (usually test scores in psychometrics) to a smaller number of underlying factors. These are then graphically rotated to produce the mathematically simplest structure. **Orthogonal Rotation** of the factor axes forces the analysis to produce independent factors. **Oblique Rotation** allows non-independent factors. Most psychometricians today use oblique rotation.

Heritability—A statistic used in Behavior Genetics that ranges from 0.00 to +1.00 that measures the percentage of the observed variation (**Variance**) that can be explained by genetic (as opposed to environmental) sources. A heritability of 0.00 means that none of the variation in test scores is attributable to genetic sources; +1.00 means that all the variation is attributable to genetic sources. Heritability is most often determined by kinship studies (most forcefully by comparing genetically identical twins reared apart against unrelated children reared together).

The **Heritability Estimate** is specific to the test, the group, and the time in which it was obtained. Theoretically, even if the heritability of test scores in each of two groups being studied were 100%, the mean difference observed *between* the groups could be entirely the result of environmental factors. Most modern behavior genetic studies yield estimates of not only the genetic and the environmental sources of variation, but also of the genetic-environmental covariation and of the genetic-environmental interaction (that is, non-additive variation).

No one claims that IQ is determined entirely by genes or by the environment. The debate centers around the amount of each. Herrnstein and Murray give a heritability range for IQ of 40% to 80%.

Psychometrics— That branch of psychology that deals with the construction and interpretation of tests that measure the variation of mental ability (most frequently IQ tests) and personality. The principal tools of psychometrics are correlational (and regression) analysis and factor analysis. **Behavior Genetics** analysis allows us to determine that amount of the variation in psychometric test score variation that can be explained in terms of genetic versus environmental sources. The principle tool of Behavior Genetics is **Analysis of Variance** (ANOVA).

Regression—The line that graphs the trend of one variable, on average (e.g., college GPA or income) against changes in a second variable (e.g., IQ scores). In **Logistic Regression Analysis**, the dependent variable is **Dichotomous** (that is, it is measured in terms of a Yes-or-No scoring system, such as being above or being below the poverty line, or being either employed or unemployed); the independent variable is continuous (e.g., income or IQ). A **Scatter Diagram** presents the complete plot of the scores obtained by each subject; the score on one variable fixes the point on the horizontal, or X-axis; the score on the other variable, the vertical, or Y-axis.

Reliability—The consistency of measurement, quantified by the correlation between test scores obtained at one time and test scores obtained by the same subjects at a different time. The reliability of IQ test scores tends to be very high. That is, individuals tend to score nearly the same on IQ tests taken at different times. **Validity** measures how successfully test scores predict (that is, correlate with) some independent measure, termed a **Criterion**. For example, you could use IQ scores to predict college GPA or simply whether or not a student graduates from college (the latter is a **Dichotomous Variable** – see definition under **Regression**).

Variance— This is a measure of the dispersion or scatter of the test scores for an entire group of subjects. Most psychometricians use the square root of the variance (termed the **Standard Deviation**) as a unit of measurement of how much the score for an individual differs from the mean for the group (or mean of one group differs from the mean for some reference group).

CHAPTER 11

ROBERT STERNBERG ON THE BELL CURVE

CONFUSION AND MISCONCEPTIONS ABOUT
THE RELATIONSHIP BETWEEN
HERITABILITY AND ENVIRONMENT

Robert Sternberg's history with intelligence testing has been described as a lifelong love-hate affair. In the 6th grade, suffering from test anxiety, he performed so poorly on a standardized test the school authorities made him take the test a second time with a group of 5th graders. Feeling confident among that group, he did quite well. (He describes the incident and its effect on him and his subsequent research in more detail in the interview.) By the 7th grade he had developed his own intelligence test dubbed the STOMA – the Sternberg Test of Mental Ability (he was certainly not lacking in initiative). As an undergraduate, Sternberg majored in psychology at Yale, and managed to get only a "C" in his introductory psych course. Once again he proved that such early predictors are not all they are cracked up to be. By the time he was a graduate student in psychology at Stanford, his doctoral work earned him the Sidney Siegel Memorial Award. He has worked at the Admissions Office and the Institutional Research Office at Yale University, the Test Division of the Psychological Corporation, and at the Educational Testing Service (ETS).

Sternberg's early work built on the standard psychometric conception of intelligence as a single, general trait (Spearman's *g*). His Componential Theory broke *g* down into its underlying information processing components. But Sternberg found that even his Componential Theory and the tests he developed to measure the component processes still missed a lot. Individuals who scored highly on Sternberg's early test still were not guaranteed success and many individuals who did not score as well went on to have a better record of real life accomplishments than did those who scored well on his or on other traditional tests.

Sternberg has moved beyond the Componential Theory to what is now known as the Triarchic Theory of Intelligence. In his view, the Triarchic Theory does not disprove either *g* or his earlier Componential Theory, but rather subsumes them under a larger framework.

The Triarchic Theory posits three facets that make up what we call intelligence (and for Sternberg that term, when properly defined and measured, must translate into real life success). The three facets are: (1)

Analytical Intelligence, which is similar to the standard psychometric definition of intelligence and corresponds to his earlier Componential Intelligence. It is measured by analogies and puzzles and reflects how an individual relates to his internal world; (2) Creative Intelligence which involves insight, synthesis, and the ability to react to novel stimuli and situations. This is the Experiential aspect of intelligence and reflects how an individual connects the internal world to external reality; (3) Practical Intelligence, which involves the ability to grasp, understand, and solve real life problems in the everyday jungle of life. This is the contextual aspect of intelligence, and reflects how the individual relates to the external world about him. In short, practical intelligence is "street smarts."

The Sternberg Multidimensional Abilities Test measures all three (on separate scales). All three are important to success in life and have been used to develop programs for children and to select business managers. Together, the three measures provided more information than just the analytical intelligence measured by standard IQ tests on which, in Sternberg's view, our society has placed far too much emphasis.

Skeptic went to Professor Sternberg to get his view of the controversial book *The Bell Curve* (see the interview with its co-author Charles Murray in the previous chapter). Having first discovered, then grasped, and for years fondled its trunk, Sternberg feels that the standard psychometric interpretation (on which so much of *The Bell Curve* is based) has mistaken the elephant of intelligence for nothing more than a big and powerful snake.

Sternberg also strongly believes that all three aspects of intelligence can not only be measured, but also developed. While he sees this technology as still being in the infant stage, Sternberg believes it (and not just measurement and selection) is the goal on which society should focus.

In 1981, Sternberg received the Distinguished Scientist Award for Early Contribution of Psychology. The citation read, in part, "He has put intelligence into investigations of intellectual abilities...combining experimental methods and theories of cognitive psychology with traditional mental-testing ideas in analyzing intelligent performance and individual differences...[and has] cross-fertilized and infused vitality into studies of individual differences and the experimental analysis of intellectual performances." He is currently IBM Professor of Education and Psychology at Yale and the author of *The Triarchic Mind*, and *Intelligence, Information Processing and Analogical Reasoning*. He is also the editor of *The Encyclopedia of Intelligence*, and on the editorial board of *Intelligence*, the leading scientific journal in the field of intelligence and mental ability.

Like Charles Murray, Robert Sternberg thinks that the study of intelligence is vitally important to the health and survival of our society, but oh how their diagnoses and prescriptions differ – to the point of possible allegations of quackery and malpractice! Here then is what "the Street Smart Psychologist" had to say about *The Bell Curve*.

Figure 11.0: Robert Sternberg

Miele: At *Skeptic* Magazine, we are interested in examining the evidence for all claims. Herrnstein and Murray present what they consider to be the consensus of scholars working in the field of intelligence. Snyderman and Rothman (*The IQ Controversy*) present similar data and there was a statement recently in *The Wall Street Journal*. You summarized this so-called consensus, if I can paraphrase, in the following manner, "IQ exists, it's heritable, there are group differences, and all this matters." But in your review of *The Bell Curve* you say, "the lay public remains sadly uninformed. Nothing could be further from the truth." So as skeptics our question is, if there is this much disagreement, is psychology even a science in the same sense that physics is?

Sternberg: There is disagreement in physics and there is disagreement in biology. Active sciences always have disagreements within them. The general public usually doesn't get the full sense of the amount of disagreement because the information is filtered through the media and what comes out is only a portion of what's actually going on in science. Normally they would not even be aware of the disagreement in psychology except that this book (*The Bell Curve*) was written for the media. It was written with the purpose of stirring up this kind of controversy. So the general public becomes aware of these types of disagreements in psychology, but they exist in every active science.

Miele: Which then is a fairer description of your own position – (1) there is a consensus on the topics that Herrnstein and Murray discuss, or (2) yes, there is a consensus among psychometricians, but I (Robert Sternberg) and others disagree with it?

Sternberg: To the extent that there is a consensus it is certainly not Herrnstein's and Murray's. You mentioned the statement that appeared in *The Wall Street Journal*, which a number of psychometricians signed. This statement was not totally coincident with the views of Herrnstein and Murray. It certainly wasn't coincident with mine. I would say that I don't think that consensus in science makes much difference. Science isn't done by majority rule. There is a misperception on the part of the public that even if you took a vote and 51% of the scientists said, "I think this is true," that would have any impact on whether it's really true or not. Science is not politics. There could be one person who believes something and that person might be right. In fact, most of the really important work in science has been the result of one person saying, "You guys are all wet, you're full of it!" and then proceeding to show that he is indeed right. So I think the conception of science as taking a vote, and whichever side gets 51% is right, is simply wrong. Science is not politics, we're not electing anyone.

Miele: OK. Let's go point by point through the earlier paraphrase of what's in *The Bell Curve*. Let's take the concept of general mental ability, which Herrnstein and Murray invoke, versus Gardner's "Seven Intelligences" or your own "Triarchic Mind," which uses the concepts of Analytical Intelligence, Creative Intelligence, and Practical Intelligence. You once wrote in an article that, "We interpret the preponderance of evidence as overwhelmingly supporting the existence of some kind of general factor in human intelligence. Indeed, we are unable to find any convincing evidence at all that militates against this view." So how then is your approach different than the approach of the *g*-theorists that Herrnstein and Murray invoke?

Sternberg: That article was published in 1983, I believe, and was based on work that I had done in the late 70s and early 80s. What I found at that time was that if you use the kinds of tasks that are used in intelligence tests, then you will get the *g* factor. That statement reflected analyses we did that instead of using individual difference analysis used process analysis. Even using process analysis, we got a general factor. So if you were to ask me, "Do I think that there is general factor in the kinds of tests that psychometricians use?" I would say "Yes." That is a different question from, "If you define intelligence, not just as IQ, but as involving more than what the IQ tests in fact test, is there then a general factor?" then I would say the answer is "No." So the way psychometricians operationalize it, you get a *g* factor. But I think, as do many other people, that is a narrow view of intelligence.

Miele: Would it be fair to say that your view is not that *g* does not exist, but that it is one facet or one side of the picture?

Sternberg: That's right, *g* is not quite as general as some psychometricians make it out to be.

Miele: Let's turn to the heritability of IQ. Let's just look at that, as you described it, narrow measure we call IQ score, within the White population. Herrnstein and Murray repeat the same numbers that we find in reviews of the literature by Erlenmeyer-Kimling and Jarvik, Bouchard and others. Do you have any disagreement with or criticism of their estimate of .40 to .80 as the heritability of IQ score within the white population?

Sternberg: I think that there is definitely some heritability of intelligence in the White population. Almost every psychologist believes there is some heritability of IQ and I agree. But the public may not understand just what that means. If you accept the use of the heritability statistic, about .5 is probably right.

Miele: Can you explain wherein the general public's conception or the media's description of what is meant by heritability is wrong?

Sternberg: The commonplace understanding of heritability often doesn't realize that heritability is calculated within a range of environments, at a given time, for a given population. So heritability is not the same in every population. In fact there is wide variation in populations over time and space. It is not a fixed statistic. The value you obtain (for heritability) depends on the population, where it is, and when it is. But the major misunderstanding relates to the role of the environment and to the role of teachability. With respect to teachability, even if heritability is fairly high, it does not mean that we cannot modify intelligence.

Miele: What exactly do you mean by that?

Sternberg: What I mean is that there is absolutely no relation between how heritable something is and the existence of a difference in group means. The most common example is height. Height has a heritability of greater than .9, but heights have increased quite dramatically in some countries like Japan and have also increased in our own country over the course of several generations. So despite the much higher heritability of height than anyone believes of intelligence, we see that height can increase. To take a more extreme example: there is a disease known as Phenylketonuria (PKU), which is 100% heritable and yet through an environmental intervention, namely withholding Phenylalanine from the diets of infants from birth, you can either reduce or eliminate the mental retardation that normally results. In other words, even when heritability is 1.00, environmental interventions still matter. There are different ways to look at intelligence. One is to do heritability statistics, which I've never found to be that helpful. Another way is to look at studies on intervention. For example, Dennis did a large study in Iran where he found that kids that were placed in Iranian orphanages, almost without exception, were mentally retarded, whereas the children who were quickly adopted before the age of two scored at normal levels on intelligence tests, roughly a 50-point difference in obtained IQ.

Miele: Are such results repeatable?

Sternberg: Yes. Obviously the environment of the Iranian orphanage was pretty bad and that's why you got that level of retardation. But if you look at the kinds of environments some of our least fortunate get, even in the United States, in the inner cities, they are not so hot either. Diamond performed studies on brain mass in rats and found that if you give them an enriched environment, it affects the brain, which becomes heavier and more convoluted.

Miele: How is your more elaborate view of heritability and its limitations different from what Herrnstein and Murray say in *The Bell Curve*?

Sternberg: The way that book is written is to, I think, say X on page 605 in sentence 8, with an appropriate caution, and then invite the reader to a somewhat more extreme conclusion elsewhere. So if you were to ask, "Is there anywhere in *The Bell Curve* that explains what heritability truly is?" there probably is. If you were to ask, "What inference do Herrnstein and Murray invite their readers to draw?" they go beyond what they know. For example, with regard to race differences, Herrnstein and Murray invite the reader to conclude that race differences are due to genetics, even though they have no evidence of that, and they know it.

Miele: They do review studies that deal with race differences.

Sternberg: Yes, but there is evidence that they do not review at all. There is nothing in the book that suggests that race differences are genetic. They even say that. But what they do say is that is what we would infer given the data, even though probably somewhere else, they would have one sentence to the effect that there is one study. And they don't cite a number of studies that suggest that race differences are not genetic.

Miele: Which studies don't Herrnstein and Murray cite?

Sternberg: Well, one study that they cite and distort the results of is the Scarr-Weinberg study. What Scarr-Weinberg and several people have done is look at blood groups (associated with Whiteness or Blackness), or skin color, and looked at the correlation with IQ. The typical correlation is about .15, suggesting that you are accounting for about 1-2% of the variance. And even that less than 2% could be due to the way darker versus lighter people are treated. So when you look at the studies that have been done, they counter-indicate the conclusion that Herrnstein and Murray draw.

Miele: Which then is your position on the question of race differences in IQ? We all see the 1 standard deviation difference in mean IQ if we give the tests to groups of Blacks and Whites. Is that mean difference the result of genetics, environment, both, or should we say at this point that we just don't know?

Sternberg: What we know is that almost any difference is some interaction between heredity and environment. But in terms of apportioning the difference, we have no idea. And I think that Herrnstein and Murray know that as well as do other psychologists. Like everyone else we don't like ambiguous situations, so some jump to conclusions even though I think at this point we don't have a very good idea of why we get that difference. Although we recognize that it has generally been decreasing over time.

Miele: You say that Herrnstein and Murray build their whole argument on the (often wrong) interpretations of statistics. Can you be more specific?

Sternberg: One example is taking studies that show that within-group heritabilities have nothing to do with between-group heritabilities and then insinuating that they do. Another example is the issue of causation and correlation. They know, and anyone who takes statistics knows, you can't draw any real causal conclusions from correlational data. Lots of things correlate with lots of things, IQ being one of them. To draw causal inferences from correlational data, which is what all their data are, is statistically incorrect. Another thing that many may not realize is that virtually all their data are based on one study, the National Longitudinal Study of Youth (NLSY), which was not a study that was particularly representative of the United States population.

Miele: In what sense?

Sternberg: The mean was low. I think the mean IQ in that group was around 94 and the standard deviation was not 15 or 16. It was not a typical US population. Another thing they do, in comparing correlations, is that they don't take into account the reliability and precision of the measures being used. For example, almost every measure we use is a proxy for something else. If you ask yourself, "How good is 'number of years of schooling' for measuring how much education a person has?" it's not a very good measure. Two people could each have 16 years of schooling, but if you compare somebody who was a straight A student in a really good college with someone who was a D student in a really poor college, the number of years of education will be the same but their educational attainments will be vastly different. So as a measure of how much schooling you have really had, years of education is extremely imprecise and it's not going to look very good in correlational analyses. In contrast, IQ is a pretty good measure of that narrow construct, compared to the other types of measurement that we have. That will make IQ look more powerful than the other measures because the other measures are such crude proxies for the constructs that they are trying to measure.

Miele: Do we have better measures of what a person has accomplished?

Sternberg: You could certainly measure their achievements with a variety of kinds of tests. Those kinds of tests are available. Herrnstein and Murray did not use them, but they are available. I'm not saying that they are perfect, but what I am saying is that the kinds of variables Herrnstein and Murray use, in contrast to IQ, are generally not very reliable and not very good as proxies of the constructs they are trying to measure. In terms of interpretation another thing that I found strange was that the data suggest that IQs generally have been going up (the Flynn Effect). Herrnstein and Murray cite Flynn's work and they agree with it, but they then find themselves in the awkward position of explaining why IQs are going up when (according to Herrnstein and Murray), IQs should have been going down. They then use very weak arguments to try to dismiss the Flynn Effect, after making a big deal about its existence. In other words, having committed themselves to a position, if the data are consistent with their position, they cite them, and if the data are inconsistent, they try to come up with what I think are fairly hare-brained interpretations of how that could happen.

Miele: You refer in your review to Binet's "Mental Orthopedics" and also to the work of Lev Vygotsky. These seem to be part of a different paradigm that sees intelligence as something that grows or forms in a social setting, and that sort of interpretation makes sense to most of us. You even mention that Herrnstein at one point developed one of the better programs for helping to raise IQ. Can you explain that further?

Sternberg: Some years back in the early 1980s the government of Venezuela initiated a country-wide drive to improve the intellectual abilities of the children. They invited a number of researchers from Venezuela and abroad to come in. One program was initiated by Harvard, and Herrnstein was the head of that program. It was successful. They published the results in *American Psychologist*, which is a leading psychological journal, showing that there had been significant and impressive gains in IQ.

Miele: How do you square that with, for example, Herman Spitz (who signed *The Wall Street Journal* statement) or even Zigler, who is one of the gurus of Head Start, who says, "It does great things for keeping children in school and out of jail, but IQ isn't something that we can really move around a lot, we just don't know how to do it." Again, there appears to be a contradiction between what the authorities say.

Sternberg: I don't think there would be that big a difference between me and Zigler. You can get some increases. I don't think we know how to get really large, long term increases. On the other hand, we're talking about work that's been around for maybe 30 years. If you ask, "how far had medicine gone in 30 years after its inception, say in Ancient Greece in terms of curing illness," it wasn't that hot either. We have just started on this kind of work. You can't expect that in a fairly short amount of time we will have figured exactly what to do. This field just isn't at that point yet.

Miele: Would you agree that the burden of proof should be on those who claim, at this point, that they can raise IQ?

Sternberg: Yes. And I think that it has been there. There have been programs, Ramey's Carolina Abecedarian Project is another, that have worked. I think it's hard to maintain the IQ gains. But if you think environment is important in the development of intelligence, and you put people in a really good program and you raise their IQ, and then take them out of the program and put them back in the poor environment in which they started, chances are you are going to lose a lot of the beneficial effect. If you give someone antibiotics for a disease, cure them, then put them back in the original septic environment, the disease will return. We've seen this when we work with children with parasitic infections. We can give them Albendazol and it will cure their parasitic infection. But if you put them back in the environment in which they acquired the infection, they will just acquire it again.

Miele: In another quote from your article, you say, "How strange that we have become a society that values what someone might do more than what he has done." Isn't that what all science is about – trying to predict in advance? Isn't the final practical test whether it works economically in the real world? Isn't that what physics, astronomy, and chemistry, all strive for?

Sternberg: No. And if you were to tell someone in physics that the final test is whether it works economically in the real world, they would tell you that that has nothing to do with truth at all. It's totally irrelevant to what is scientifically true. There is a question of what is true and then there is a question of how you use it. But I don't think any physicist would believe that the ultimate test of truth in physical science is economic exploitation or usefulness.

Miele: Do you agree that science tries to predict things?

Sternberg: Sure, one of the goals of science is prediction.

Miele: Then is your quotation a criticism of psychometrics as a science or of what we do with its results?

Sternberg: My criticism has nothing to do with scientific truth. It has a lot to do with what I see as a maladaptive oddity in our society. We often seem to value the ability to do something more than what people actually do, and there are lots of examples of that. We are one of the few societies that place so much emphasis on intelligence tests. In most societies there is more emphasis on what people accomplish. I do think there is something to be said for emphasizing what people actually do, rather than what they might or might not do. We have gotten into the somewhat ridiculous situation of even having constructs like "overachiever," which is talked about in education. Most societies wouldn't have that construct. What does it mean? An overachiever is somebody whose IQ is lower than their achievement scores. The idea is that they are achieving too much and that there is something wrong with them, and that they ought to be pushed back to their own size.

Miele: It seems logically impossible.

Sternberg: It is! So where do we get that idea? We got that idea from valuing the predictive test more than the achievement. You get a lot of people who work in admissions who will count SAT scores or IQ tests or ACT tests more than they count actual individual accomplishment. I think that that is a backward way of looking at things.

Miele: The other side of the argument came out when I interviewed Douglas Detterman, the editor of *Intelligence*. He reviewed for me the work at Brooks Air Force Base where *g* proves to be far and away the best predictor of pilot training and a whole host of other things. So getting back to the question of economic reality, the Dallas Cowboys want to know who's going to make the team before they go through training (if possible). When I was in the Air Force we had big problems with washouts from pilot training. It costs a huge amount of money to put someone through that training. So isn't it just economic good sense to use the best tests and measures for prediction?

Sternberg: I am not disagreeing that IQ is predictive of a lot of things. I'm not one of the extreme left-wingers who say that IQ tells you absolutely nothing. I don't agree with that. So to the extent that it predicts some level of success in pilot training, I don't have any argument with that. But I do argue with the idea that IQ is the end of the line. We have been working for about 10 years in the field of practical intelligence, predicting, for example, the success of managers and sales people, which are pretty practical occupations. We actually did a study at Brooks AFB and found that our measures of practical intelligence – that is, measures of how well you can go into an environment and figure out what you need to succeed in that environment and then actually do it – predict job success in managerial jobs and in sales jobs at least as well and arguably better than IQ tests. Moreover they do not correlate with IQ tests, which means that: (a) IQ is not the only predictor, and (b) the kinds of predictors we have are relatively independent of IQ. That's not to say that one is important and the other is not. Rather, it says that both are important and that there's more to predicting success than just using IQs. If you want to predict success in jobs, I'm not saying that IQ is worthless, but I am saying that it's not the only thing you can use.

Miele: Don't we find the situation in sports all the time where we say, "Hey, this guy came in with great abilities – he runs fast, and so on – but he just isn't a player, he just doesn't come through." And there are in baseball, for example, these Billy Martin-type players, who you think will never make the team, but they play their heart out on every play. When it's the World Series I know who I want in my lineup. Is that the sort of argument that you are making?

Sternberg: I think you are talking about something slightly different. Practical intelligence is different from academic intelligence, and motivation is something different again. And motivation is also very important.

Miele: But what about those things in sports like knowing the game, knowing the tricks of the trade?

Sternberg: That's what I am talking about when I speak of practical intelligence as opposed to academic intelligence.

Miele: Often the guys who aren't the best players are the best managers because they've had to learn all this in order to stay in the league.

Sternberg: And some of them are good players too. You don't need research to tell you that. We all know people who had very high scores on SATs, GREs, and IQ tests, who don't seem to be able to translate it into any kind of success in their lives. It's not to say that everyone with a high IQ is going to fail. Obviously, that's not true. The other thing is that a lot of what is said about prediction is, in my view, somewhat shaded and one can even argue falsified, by the fact that we have already been using IQ in our society in its various disguises to create the truth of what is being predicted.

Imagine that we had decided that in order for someone to succeed in college they had to be over six feet tall and so you only accepted people who were over six feet tall. Well, within a generation or two, you would find that most of the people who were in the high paying jobs were over six feet tall. And you would note a correlation between success and being over six feet tall. But why did you get that correlation? Because you created a system to make that come true. We have created the kind of system that Herrnstein and Murray talk about by using SATs for college and GREs for graduate school, LSATs for law school, and GMATs for business school. In other words, almost any access route to the high-paying occupations requires you to do well on these tests. That will artificially and spuriously create the correlation between high scores and entering into top jobs. So what Herrnstein and Murray are talking about when they describe the cognitive stratification of modern America, is, in fact, an invention of our society.

Miele: But haven't path analysis studies shown that even IQ measured at an early age is a better predictor than a host of other things measured later on?

Sternberg: Early childhood IQs do not predict anything accurately. By around the age of eight, when IQ becomes stable, you are predicting about 50% of the variation in adult IQ.

Miele: I'm referring to studies that show that childhood IQ predicts adult earnings. In other words, IQ, rather than being reared in an affluent home, led to higher adult earnings.

Sternberg: Certainly IQ will be somewhat predictive of earnings. That's not inconsistent with what I just said. If you have a high IQ, even if you come from a not very affluent home, you are going to be rewarded by our society for it. If you come from a poor home and you do well on the SATs, Yale and Harvard and Princeton will all be begging to have you so that they can claim diversity. That's the ticket to success.

Miele: Yes. And wouldn't a lot of people say that that's all for the better? That it is extending opportunity in our society beyond what it was 100 years ago, when everybody at Harvard came from a rich home?

Sternberg: If you ask me, "Is it better to use family affluence or IQ," I'd say that its better to use IQ. But if you were to ask me, "Is it better to use IQ or demonstrated accomplishments?" I would say it is better to use demonstrated accomplishments. My concern is not that there will be people with high IQs who don't do so well. That's not the problem. The bigger problem, in my view, is the people who don't test particularly well, who are going to be screwed. I've seen a lot of them, including me in my very early years of school.

Miele: Really?

Sternberg: When I was very young, I did poorly on IQ tests because I was test anxious. The result was that teachers had low expectations for me and I wanted to please my teachers. So I met their low expectations. They were happy and I was happy that they were happy. I've been there and I've seen it happen to lots of people I know. I got over my test anxiety and then did extremely well on tests. All of a sudden the expectations were high. To a large extent it becomes a self-fulfilling prophecy, either way. So when you tell me that IQ predicts later success, sure it does. You get low scores on your tests, everything starts to change in your life and you're on a downhill slide. It's not a controlled experiment, because the very score itself is having an effect on where you're going to be allowed to go.

Miele: Do you think it's just as simple as you said, or were you just citing one specific example?

Sternberg: There are many, many examples of that. Anyone who has ever worked in college or graduate school admissions, which I have, will see cases of people who don't test well and therefore aren't accepted. You can understand the constraints on them. If they let the average scores go too low, they're going to start losing prestige and their best students will go somewhere else. So there's pressure to keep the scores up and there are lots and lots of people who don't score well on those tests who can't get in anywhere. You don't even have to do a scientific study to know that. Suppose you're very creative but you don't test well. You're out of the game. (And no one is arguing that those tests measure creativity.) So, some people who are very creative or very good at practical intelligence are not allowed access and don't make it through that kind of a system. Later when they are not in such good occupations or making as high a salary, someone will point out that there is a correlation between IQ and salary and they will be right – we created it.

Miele: Let me turn briefly to Howard Gardner's "Multiple Intelligences." In *The Triarchic Mind*, you say that they should actually be referred to as talents, rather than intelligences. How is your criticism of Gardner's model different from the *g*-theorists' criticism of Gardner – namely that he is arguing for independence of things we really wouldn't call intelligences?

Sternberg: There is no absolute, agreed-on definition of "intelligence." One of the battles in the field, arguably *the* battle, even more than the heredity-environment issue, is what you include under the definition of intelligence. There's no final answer to that because God doesn't tell us what he means. To a large extent, intelligence is our own creation. It is a creation to describe the fact that in terms of adaptive skills, some people have more than others. I think that you can argue that practical abilities are adaptive. We showed that they are adaptive for everybody. Everybody needs the ability to be able to work with other people, to be able to figure out the environment that they are in. It doesn't matter who you are. I'm supposedly in the ivory tower, but if you don't have practical ability, you don't stay here very long. You also need creative intelligence in the vast majority of jobs and in your life, because you can't always use solutions that you or someone else came up with before. Life just isn't that way, especially in very quickly changing times such as those in which we live. To me, something is a part of intelligence if it's necessary for adaptation. What I argue is that something like musical talent – or musical intelligence – is not in the same class as practical intelligence. A talent to me is something that is important in the lives of some people. If you want to go into music, musical talent or musical intelligence is important. But for my career even if I was totally tone deaf and never played any music at all, it wouldn't make any difference to my life or the life of anyone else.

Miele: This is very similar to what Charles Murray told me when I interviewed him and asked him about Gardner's work. He said, "*g* is the only one that if you do badly enough on it, you end up in a home for the retarded."

Sternberg: I don't think it is true that *g* is the only one that if you do badly, you are screwed in life.

Miele: What other ones are there?

Sternberg: I just told you. You get people who have very high *g* who may not end up in a home for the retarded, but their lives don't differ a whole lot. They keep screwing up everything they do. And I've seen them in my own field. People who are very bitter because they can't get along with anyone. All of a sudden they find that no one is inviting them to give talks, they don't stay very long in a job, no one wants their articles, they don't know how to communicate, they don't know how to argue for a position effectively, and all the *g* in the world doesn't save them.

Miele: Some would say that what you're focusing on are really personality measures.

Sternberg: Personality is very important for life too, but it's different from our tests of tacit knowledge. The tests of tacit knowledge that we use to measure practical intelligence measure what you know. They do not measure your personality. Introversion and extroversion are personality measures. The difference is that you can be very successful in life as an introvert or an extrovert. You can be successful in life being somewhat altruistic or not very altruistic. You can't be successful in life if you lack practical intelligence.

Miele: We live in the computer and information age and although there are individual differences in speed of learning, so what if it takes John 100 trials to learn the alphabet but Mary only needs 15? When we were in school, the kid who took longer or who couldn't throw a baseball as far was ridiculed and embarrassed. But computer-assisted, self-paced learning systems let you go at your own rate. There's lots of time in childhood, so what's the problem?

Sternberg: You're arguing with the wrong guy. I agree with you. I've made the same point in some of my writings. Our society is a little bit unusual in putting so much emphasis on speed.

Miele: It seems to me that a lot of the things that you are complaining about are the result of a social setting in which we teach people instead of letting them go at their own rate. If they make a mistake, they can try it again, but no one gets laughed at. So how much of your own work has gone towards developing computer-assisted training systems?

Sternberg: I do a lot of work on training systems, but for the most part they are not computerized. But the characteristics you're talking about are not limited to computerized systems. You're talking about values, not the necessity of using the computer as a medium. I think the computer is a good medium, but there are other ones as well.

Miele: Turning to your own work, one of the criticisms would be, "What have you added beyond what's in, say, John B. Carroll's extensive factor analysis in the domain of abilities?" Or are we going back to the earlier question in which you said that the practical intelligence is an equal ability with *g*, and not just a personality factor or "tricks of the trade" or just accumulated job knowledge? Are your tests in fact just job knowledge, rather than a continuing capacity of the individual?

Sternberg: That criticism is simply false. If you look at the correlation between tacit knowledge for being an academic researcher and tacit knowledge for being an executive, the correlation was pretty high – about .5 to .6. But in terms of teaching job knowledge, probably no one who is a psychologist went to business school or vice versa. Tacit knowledge is something you pick up from the environment. I don't care what you call it. If you want to call it job knowledge or the ability to use job knowledge or the ability to use common sense knowledge that you pick up, the name isn't important. What I'm saying is, whatever you want to call it, it's at least as important as the academic sort of intelligence and it's not the same thing as that. I don't need to argue about the name attached to it.

Miele: Let me turn this around. Suppose that I do great on the SAT and the GRE and the IQ tests, but I do terribly on Sternberg's practical knowledge tests. Does that mean that I'm not going to be allowed to go to graduate school, and instead just sit home and watch quiz shows? What's your remedy or treatment for that?

Sternberg: Well, I'll tell you what we did. In the program that we have run for the past two summers here, based on my own model, students were given my abilities test (which had analytical, creative, and practical sections), and we admitted kids in the experimental groups in one of four ways – either very high analytic, very high creative, very high practical, or balanced. In other words, the model is that very few people are going to be good in everything. I wasn't interested in taking an average. Rather, what I was interested in is the fact that people have different patterns of strength. What you want to do is to help them capitalize on whatever their pattern of strength is. So, if you're very good in analytic, not so high in practical, that's fine. That's an important kind of skill too. Analytical skills should not have absolute preference over creative or practical skills. I'm not saying that these other skills should have absolute preference over analytical skills.

Miele: Another criticism of your work – I think this was stated by Messick of the Educational Testing Service – is that you are actually invoking more explanatory entities than you are measuring, thereby violating Occam's Razor. In other words, you have come up with fewer test measures than entities you have invoked to explain the measurements you have?

Sternberg: I've talked about analytical, creative, and practical intelligence and I have measures of all three. If you look at the Triarchic Theory and you look at all the constructs, for example, metacomponents (higher order processes) like defining problems, or setting up strategies to solve problems and you ask if I have a measure that isolates every single one of those, I don't. Nor do I particularly want to do what J. P. Guilford did in his career, which was to spend his whole career filling in the blanks. That's the sign of an uncreative career, at least in my view. If we talk about the basic structure of the theory and the three parts of it, I have measures for all of these, we've used them, and they work. The fact that Messick works at the Educational Testing Service might give him a certain vested interest in what ETS does.

Frank Miele

Miele: One of the criticisms of *The Bell Curve*, especially the now infamous Chapter 13 on race differences that the popular press makes a lot of, is that Herrnstein and Murray have cited the so-called "Tainted Sources," particularly people who have been funded by the Pioneer Fund. But some of the individuals funded by the Pioneer Fund have articles in your own *Encyclopedia of Intelligence*. Do you want to comment on their work?

Sternberg: I have been offered funds by an organization and I didn't take the funding because I didn't like the organization. That was my personal choice. If they funded my work, the fact that they funded it doesn't mean that the work is therefore invalid. Work stands or falls on its own, regardless of who funded it. For example, Messick is at ETS, so I tend to be somewhat skeptical of people at ETS. But that doesn't mean that because they're at ETS, what they say is wrong. It just makes me think twice about what they say.

Miele: The same thing could be turned around said about Sternberg or Gardner. They have their vested interests.

Sternberg: Absolutely. I agree with that. If you are coming from the other point of view, you ought to say that I have a vested interest. That's why I say arguments stand or fall on their own.

Miele: In the *Encyclopedia of Intelligence* you do have articles and I think even biographies of Arthur Jensen and Hans Eysenck and others. So you must have some respect for the work they have done, at least in specific areas.

Sternberg: Even though I may not agree with a lot of what Jensen has said, the importance of a person's contribution is not determined by how much I agree with what they say. I think that he's made a contribution. His work on race I utterly don't like, but I don't think that that's where his scientific contribution lies. I think his contribution is the work he's done on reaction time and intelligence.

Miele: If I can summarize, it seems that your main interest is in the nature of intelligence rather than in the differences between individuals and groups. Is that right?

Sternberg: I'm interested in both. I just think that what has happened is that a small part of the question has been treated as if it is the entire story and it's not.

CHAPTER 12

IQ IN REVIEW

GETTING AT THE HYPHEN IN THE NATURE-NURTURE DEBATE

REVIEWS OF:

Gottfredson, L. (Guest Editor) 1997. "Intelligence and Social Policy," a Special Issue of the Journal *Intelligence*. Vol. 24, No., January-February 1997. pp. 320. ISBN: 0160-2896.

Nyborg, H. (Editor.) 1997. *The Scientific Study of Human Nature: A Tribute to Hans Eysenck at Eighty*. New York: Pergamon. pp. 621. ISBN: 0-08-042787-1.

Segal, N., Weisfeld, G., and Weisfeld, C. (Editors) 1997. *Uniting Psychology and Biology: Integrative Perspectives on Human Development*. Wash., D.C.: American Psychological Association. pp. 568. ISBN: 1-55798-428-X.

Sternberg, R. and Grigorenko, E. (Editors), 1997. *Intelligence, Heredity, and Environment*. NY: Cambridge University Press. pp. 608. ISBN 0-521-46904-X.

Sternberg, R. (Guest Editor), 1997. "Intelligence and Lifelong Learning," a Special Issue of the Journal *American Psychologist*. Vol. 52, No. 10., October 1997. pp. 320. ISBN: 1-55798-469-7.

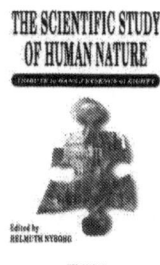

Figure 12.0

Each of these publications consists of articles by a number of authors who are specialists within their respective fields. Merely listing all of them would exceed the word count allocated for this review. Rather than providing a comprehensive and evaluative summary of each article, this review will call attention to the most important contributions, explain what these volumes have in common, where and how they differ, and why they merit a review in *Skeptic*. While they are written for a professional audience, there is much that interested general readers will find of value in these volumes, particularly because it differs both in content and in tone from what they might conclude after reading only the popular or even the skeptical literature.

Two volumes, *Uniting Psychology and Biology* and *The Scientific Study of Human Nature,* are Festschrifts, written to honor two of the most important figures in contemporary psychology – Daniel G. Freedman and the late Hans J. Eysenck (who died between the time the Fest was held and the publication of the companion volume). They are important both for what these two giants contributed to modern scientific psychology, and for demonstrating how their research strategies and outlooks differed. Two other volumes, "Intelligence and Social Policy" and "Intelligence and Lifelong Learning," are special issues of top professional journals (*Intelligence* and *American Psychologist*). The former contains articles by experts who more or less support much of the scientific, though not necessarily the policy, conclusions of *The Bell Curve*. The latter was apparently brought in response by those who disagree. The final volume, *Intelligence, Heredity, and Environment*, is in the form of an extended debate between the two sides. It concludes with two assessments of which side won the debate and provides an appropriate place to conclude this review.

Frank Miele

THE FREEDMAN FESTSCHRIFT

Daniel G. Freedman, emeritus professor of psychology at the University of Chicago, has made major contributions to child development, behavioral genetics, ethology, and evolutionary psychology. Author of three books and over 50 articles, Freedman pioneered the use of film in recording studies of human behavior, combining the methods of ethological observation with those of controlled experimentation. His research and that of his many students have played a major role in establishing the extent to which behavior is best understood in terms of evolution, genetics, and development, rather than simply learning and culture. In particular, Freedman and his colleagues have explored the "hyphen in the nature-nurture problem" – the way the genome guides the individual to select particular environments during the course of development.

Freedman's dissertation, later published in the prestigious journal *Science*, studied the effects of permissive versus stern training on four breeds of dogs – basenjis, beagles, shelties, and wire-haired fox terriers. The different training styles had different effects on the beagles and the fox terriers (both breeds being highly attracted to the trainer), but not on the shelties (who are rather inhibited) and the basenjis (who are somewhat aloof).

Another classic paper showed that children born blind respond to their mothers' voices by "looking" at the source of the sound, as if they were seeing them. Freedman interpreted his results not only as an expression of an inborn orienting response, but as an evolved mechanism that enhances the process of mother-infant attachment. The most controversial of Freedman's classic papers, written with his wife, documented differences in movement between Chinese-American and European-American newborns measured immediately after birth and into the first months of life. European-American newborns moved significantly more and earlier than did Chinese-American newborns. Subsequent research confirmed this finding and showed that Navajo newborns behave more like Chinese-American than European-American newborns, thereby agreeing with other evidence for the Asian origin of Native Americans. Submitted to *Science*, the paper on racial differences in neonate behavior was rejected on a split vote of the reviewers. Freedman wrote the editor saying that it was more significant than the two papers he had already published in *Science* and requested a second review, which again rejected the paper following a split vote. Freedman then submitted it to *Nature* (the British analog to *Science*), where it yet again drew a split vote, which the *Nature* editor broke by voting for publication (p. 459). (For a popular account of Freedman's work, see his *Human Sociobiology: A Holistic Approach*, NY: Free Press, 1979).

Three contributions in the Festschrift deserve special mention. J. Michael Bailey's "Are Genetically Based Individual Differences Compatible With Species-Wide Adaptations?" addresses one of the most important debates raging entirely on the nature side of the hyphen – the different assumptions and research strategies of evolutionary psychology and of behavior genetics. While all too often viewed by pure nurturists as one in the same, they are, in fact, the respective intellectual progeny of Charles Darwin and his cousin Sir Francis Galton, and rarely overlap. Bailey's computer literature search of the Medline database for the last decade showed that while the keywords – evolution, genetics, behavior, and human – each referenced thousands of articles, only one article referenced all four! (p.82). Evolutionary psychology seeks to explain behavior in terms of species-wide universals that were selected over the course of evolutionary history. Behavior genetics, on the other hand, is concerned with the differences between individuals, and groups, within a species. (For more, see Chapter 1). So how can genetic variation persist in the face of species-wide selection pressure to reduce it? If opposable thumbs gave humans an evolutionary advantage, why should there be any difference in size, shape, or dexterity of thumbs? (There's some, but not much.) And if there was selection pressure for bigger brains, why should there be variation in the brain size in living humans. (There's a good bit, but what does it mean?) Bailey also points out that heritability estimates are a starting point, not an ending point to the study of behavior. The task ahead is to map the genotype-to-phenotype pathway by which various behaviors develop.

While the twin study, first devised by Galton, is one of the classic techniques behavior geneticists use to determine heritability (the percentage of the observed variation attributable to genetic sources), Nancy Segal's article, "Twin Research Perspective on Human Development" follows Freedman's lead and looks at twins to see how different degrees of genetic relatedness influence social interaction, comparing grief intensity for deceased twins (both identical and non-identical) as opposed to that for other relatives. Segal also compares the similarity of identical and non-identical twins to adoptive children of the same age reared together since infancy (who thus mirror the environmental similarities of twins, with none of their genetic similarity). The correlation in IQ for the unrelated sibs reared together was 0.17, well below the correlations for identical twins reared together (.86), identical twins reared apart (0.72), non-identical twins reared together (0.50 - 0.60), and ordinary siblings (0.47).

Robert Trivers first introduced the concepts of reciprocal altruism, parental investment, and parent-offspring conflict. In his chapter, "Genetic Basis of Intrapsychic Conflict," he reviews the evidence that questions one of the most basic assumptions of genetic research – that a gene is a gene is a gene, whether inherited from the mother or from the father. We now know that a small number of genes are expressed in an individual only if inherited from a particular parent (known as genomic imprinting, not to be confused with the ethological concept of behavioral imprinting). While most genes have only a 50-50 probability of coming from a particular parent, imprinted genes have a 100% certainty of coming from the particular parent. Because of the different evolutionary "agendas" of males and females, imprinted genes in the same individual can be in conflict with each other. Trivers reviews the evidence that suggests that the neocortex and most of the brain are more maternal in genetic origin, while the hypothalamus (seat of basic drives such as hunger, thirst, and some aspects of sexuality) is paternal. So the "family feud" may actually take place within each one of us. Trivers concludes that, "the neocortex might be saying metaphorically, 'Family is nice, family is important, I like family,' whereas the hypothalamus might answer, 'Me first!'" (p.393).

THE EYSENCK FESTSCHRIFT

The late Hans J. Eysenck was one of the most important, controversial, and prolific figures in modern psychology. Eysenck's citations in the Social Science Citation Index are exceeded only by those of Karl Marx and Sigmund Freud (p. 543). Eysenck, who died on September 4, 1997, two months after the Fest was held, made major contributions to replacing "depth" psychotherapies (such as psychoanalysis) with more behavioral methods, and to the study of personality, and intelligence. His most controversial work was on race and IQ and on parapsychology, both of which are discussed by the contributors to this volume.

Following in the "London School" of Galton and Burt at the University of London Institute of Psychiatry, Eysenck developed the three factor theory of personality (the factors being introversion-extroversion, neuroticism, and psychoticism), at a time when most of the social sciences rejected the notion of stable trait theories of personality in favor of the theory of situation-dependent behavior. Today, trait theories of personality are back stronger than ever, the main argument being between Eysenck's "Big Three" and the very popular "Big Five," (which splits psychoticism into dimensions of agreeableness and conscientiousness, and adds the factor of openness to experience into the mix). The number and structure of personality factors, their heritability, and their relationship to sexual behavior, crime, and genius are discussed by a number of contributors to this volume.

Arthur Jensen discusses Eysenck's contributions to the study of IQ in his chapter, "The Psychometrics of Intelligence." Jensen demonstrates that the *g* factor "is not an artifact of test construction or item selection, as some test critics mistakenly believe" and that in fact, "it is empirically impossible to devise a cognitive test that has nonzero variance and in which there are statistically significant negative interitem correlations" (p.223). In layman's terms, this means that any test that successfully distinguishes between those who perform differently on a real-life criterion always shows a g factor. Despite the claims of testing critics, says Jensen, the "imaginary test for which the average interitem correlation is either negative or zero is the psychometric equivalent of a perpetual motion machine" (p.223). Jensen reviews the evidence for the biological basis of g, such as its correlation with reaction time tasks so simple subjects can't even tell how well they are doing, its correlation with evoked brain potentials, nerve conduction velocity, and glucose metabolism in localized regions of the brain, and concludes that g is "the only 'factor of the mind' that cannot possibly be described in terms of any particular kind of knowledge or skill, or any other characteristics of psychometric tests. The fact that psychometric g is highly heritable and has many physical and brain correlates means that it is not a property of the tests per se. Rather, g is a property of the brain that is reflected in observed individual differences in the many types of behavior commonly referred to as 'cognitive ability' or 'intelligence'" (p.234). (The definitive statement of the evidence for g, Arthur Jensen's tome, *The g Factor*, has just been published by Praeger).

Of special interest to the readers of *Skeptic*, though only tangentially related to the nature-nurture debate (after all, psi-ability, if it existed, could be hereditary), is the chapter by G. A. Dean, D. K. B. Nias, and C. C. French, "Graphology, Astrology, and Parapsychology." Consider these two statements:

> Unless there is a gigantic conspiracy involving some thirty University departments all over the world, and several hundred highly respected scientists in various fields, many of them originally hostile to the claims of the psychical researchers, the only conclusion the unbiased observer can come to must be that there does exist a number of people who obtain knowledge existing either in other people's minds, or in the outer world, by means as yet unknown to science. (p.528)

And:

> Investigators who cannot explain every trick performed by stage magicians should consider themselves barred from investigating alleged psi phenomena. (quoted on p. 532).

Those familiar with the psi and skeptical literature will have little trouble guessing that the first comes from Eysenck, whose pop book, *Know*

Your Own Psi, the Amazing Randi reviewed as "a disaster in every way except one: it may provide us with an accurate picture of just how naive the authors are in designing the proper protocol for testing psi-powers" (quoted on p. 532). The second quotation is not from Randi, but, surprisingly, from Eysenck's own article on "Theories of Parapsychological Phenomena" in the 15th edition of the *Encyclopedia Britannica* (1974). Dean, Nias, and French provide a careful review of Eysenck's on-again-off-again relationship with fringe science, which is perhaps best explained by his statement, "it is said one should not waste time on topics which are obviously absurd.... I do not believe that a priori judgments of this kind are admissible in science; scientists have been wrong too many times in making explicit statements of this kind to be considered infallible. In any case, the time that is wasted is mine, and to waste it by reading the literature on astrology and parapsychology is probably better spent than in watching pornographic films, or becoming a football hooligan!" (quoted on p. 512). Maybe – but others of us might find eroticism and sports welcome diversions, allowing us to recharge our over-taxed *g* factor for later use on more productive topics. More seriously, Dean, Nias, and French effectively use the new statistical tools of meta-analysis and effect size to conclude that:

> (1) It is easy to be impressed by statistical significance. But if sample sizes are large, the corresponding effect sizes can be tiny. (2) It does not take much in the way of flaws (errors, biases, etc.) to produce tiny effect sizes. It is arguable whether we have enough experience of flaws at this level to be confident of their absence. (3) Only if the effect is repeatable do we have something that can be properly investigated. (p.536).

THE SPECIAL ISSUE OF *INTELLIGENCE*

Edited by Linda Gottfredson, this special issue was "planned as an informative extension of the collective statement, 'Mainstream Science on Intelligence'." That statement, largely in support of the scientific arguments in *The Bell Curve*, published in *The Wall Street Journal* of December 13, 1994, and reprinted in this volume, was issued to correct "opinions about human intelligence that misstate current scientific evidence" and point out that "some conclusions dismissed in the media as discredited are actually firmly supported" (p.13). These conclusions are summarized by John B. Carroll, the world's leading expert on factor analytic studies of intelligence, in his article, "Psychometrics, Intelligence, and Public Policy," which seeks to correct the fact that many "public intellectuals" in reacting to *The Bell Curve* have left the public the impression "psychometric research on cognitive ability is a discredited pseudoscience alien to the ideals of democracy" (p.26). Leaving aside the controversial issue of whether there is a genetic factor in the mean differences between groups, Carroll concludes that Herrnstein and Murray were correct in stating in *The Bell Curve* that:

> (1) psychometrics (literally mental measurement) is a rigorous scientific discipline that has resolved many questions concerning cognitive abilities; (2) general ability scores should be taken not as direct measures of hereditary intelligence, but rather as measures of rate of progress over the life span in achieving full mental development; (3) there are many other cognitive abilities besides g; (4) important sources of variation in g or IQ are environmental; (5) the IQ is possibly more an indicator of how fast the individual can learn than it is of the individual's capability of learning; and (5) much more research is needed to resolve questions about the role of cognitive abilities in a democratic society.

Psychometrically sophisticated critics such as Robert Sternberg (editor of the special issue of *American Psychologist*), Carroll concludes, are justified in stating that none of these conclusions "is actually beyond technical dispute because it is possible to cite statements by various investigators to the effect that one or another of these propositions can be questioned on the basis of relevant theory, research, and analysis," only in the sense that "almost nothing in science (at least in the social sciences) is beyond technical dispute" (p.26).

In "A Place at the Policy Table? Behavior Genetics and Estimates of Family Environmental Effects," David Rowe explains how behavior genetic methods actually are necessary to identify which of the many environmental influences actually affect cognitive ability and personality. One of the major findings is that within-family differences (those that affect one sibling and not another) are more important than between-family differences (those that affect all members of the family). It is precisely these within-family effects and their relationship to an evolutionary psychological understanding of sibling rivalry that Frank Sulloway has used in explaining the effects of birth order on personality and scientific creativity (*Born to Rebel: Birth Order and Creative Lives*. Pantheon, 1996). Rowe points out that within family effects are the least modifiable by government-sponsored programs and this needs to be understood in shaping successful public policy.

Critics of the inheritance of mental ability who routinely claim that "no one has ever identified a gene for intelligence" need to read Robert Plomin and Stephen Petrill's paper, "Genetics and Intelligence: What's New?" which explains how state-of-the-art research is moving beyond the nature-nurture question "to harness the power of molecular genetics to identify the specific genes responsible for the substantial influence of genetics on intelligence" (p.66). The genes for phenyketonuria (PKU) and for Fragile-X Syndrome that produce mental retardation have already been identified. The new technology of Quantitative Trait Loci (QTL) has identified the gene for a specific protein (Apo-E) so that "the single most important risk factor known for Alzheimer's disease can now be assessed from a few drops of blood" (p.70). This same technology allows us "to use DNA markers to identify not the gene for intelligence, but some of the many genes, each of which makes a small contribution" (p.68). While still in the early stage of comparing identified and replication groups of high and low IQ for specific genetic markers, the IQ QTL Project has found 100 DNA markers in or near relevant genes that may account for about 2% of the IQ variance in the general population.

THE SPECIAL ISSUE OF
AMERICAN PSYCHOLOGIST

Edited by Robert Sternberg, this special issue is "devoted to the exploration of the role on intelligence in lifelong learning" and making clear "the many interfaces between intelligence and learning throughout the life span" (p.1029). Its contributors include Sternberg (see Chapters 10 and 11). In line with Carroll's conclusion in the special issue of *Intelligence*, brought out by those generally in support of the scientific conclusions of *The Bell Curve*, the special issue of *American Psychologist*, brought out by those generally critical of that book, does not consist of debunking and refutation of any alleged pseudoscience, but rather of qualifications, caveats, and concessions to the general interactionist point of view that sees both heredity and environment as inextricably bound in the development of behavior. Halpern's article, "Sex Differences in Intelligence," applies this perspective to the controversial topic of sex differences. The weight of the evidence shows that "females, on average, score higher on…long-term memory, production and comprehension of complex prose, fine motor skills, and perceptual speed," while "males, on average, score higher on…spatiotemporal responding and fluid reasoning, especially in abstract mathematical and scientific domains" (p.1091). In "Intelligence, Training, and Employment," Richard Wagner tells readers that "after 85 years of research, cognitive ability tests are among the most reliable measures available to social scientists" (p.1059); Douglas Detterman and Lee Anne Thompson conclude that for IQ intervention programs to have "even limited effectiveness [they] must be pervasive, intense, and sustained" (p.1087); Liza Suzuki and Robert Valencia in "Race – Ethnicity and Measured Intelligence" state that "numerous research studies inform us…that racial-ethnic differences in the United States are among the most thoroughly documented findings in psychology" (p.1104); Sternberg in his conclusion to the special issue tells readers that "the positive manifold, whereby virtually all traditional cognitive-ability tests intercorrelate with each other, is as much an established fact as is any other in the literature of psychology. Moreover, traditional tests predict learning both in school and beyond" (p.1138).

INTELLIGENCE, HEREDITY AND ENVIRONMENT

Intelligence, Heredity and Environment, edited by Sternberg and Elena Grigorenko, includes a number of contributors from the previous four volumes (Jensen and Plomin for the hereditarians, and Sternberg and Ceci for the environmentalists, with Eysenck, interestingly, having contributed a chapter on raising IQ by vitamin and mineral supplements). Earl Hunt, professor of psychology and computer science at the University of Washington and best known for chronometric analysis of cognitive abilities, reaches similar conclusions to those above in his invited summary of the debate:

> Benjamin Franklin said that nothing is certain except death and taxes. But that was in his day. Modern-day psychologists have the Superbowl and the nature-nurture debate. The Superbowl begins with great publicity and proceeds to a lop-sided victory by one side, after which the winning players graciously tell reporters how well the losers played. So it is with the nature-nurture debate. Nature wins, 48-6, and then the winners say that, well, some of those misguided environmentalist arguments were very good tries, albeit a trifle misguided. (p.19)

Hunt's verdict is not based solely on the preponderance of evidence, but also is a direct consequence of what makes the scientific method effective – its advancing and self-correcting nature.

Neither world views that emphasize individual malleability nor views that emphasize genetic determination are wrong. They both have their place and they both can lead to good scientific analyses. The choice between the two world views determines the sort of science one does. Scientific arguments within one world view cannot be opposed by arguments for an alternative world view. On the other hand, when both world views lead to scientific analyses dealing with the same phenomenon, we can make a meaningful contrast between the two analyses, and, by implication, between the two world views, using the scientific rules for the analysis of evidence.

Even the concluding second opinion from Irwin D. Waldman, provided to balance Hunt's, does little to challenge Hunt's one liner that, "when the question is posed as Nature vs. Nurture, the debate looks like a stomping match between Godzilla and Bambi" (p.546). Waldman concedes that Hunt "speaks for the masses or at least for sizable contingents of researchers who occupy very different positions on the relevant issues" (p.552), and he repeats many of the qualifications and caveats in Sternberg's summary in the special issue of *American Psychologist*. These concern alternative models to Spearman's *g* – Sternberg's triarchic system of analytical, creative, and practical intelligence, Gardner's multiple intelligences model that includes linguistic, logical-mathematical, spatial, musical, bodily-kinesthetic, interpersonal, and intrapersonal intelligence, to which he has recently added naturalistic, spiritual, and existential intelligence, the importance of non-IQ factors in educational and vocational performance, and the role of heredity and environment in extremely impoverished groups and among non-White populations. The most controversial question of all remains the role, if any, of genetic factors in the observed mean differences between racial groups. While the data consistently show a mean group difference, enormous variation and the full range of ability can be found within all groups. The evidence for the role of hereditary factors in the mainstream, White population, does not in itself prove that the difference between groups is genetic. (For a strong statement of the hereditarian position, see Richard Lynn's article, "Geographical Variation in Intelligence" in the Eysenck Festschrift).

None of these caveats and qualifications offer an escape from hard reality of the importance of IQ in today's high-tech society and the hard choices it forces on us. Surely there are factors beyond *g*, though they may better be thought of as personality factors or other types of ability, but then what role do nature and nurture play in their development, are there individual and group differences in them, and how well can they predict educational and vocational and job success? Are college or job applicants any more likely to be satisfied if told they were rejected not because of their scholastic aptitude or IQ, but because their creative intelligence was too low, or if they were steered in another direction because of their exceptionally high musical or kinesthetic abilities? If IQ scores can be abused, the misuse of such other factors could lead to stereotyping with a vengeance. If you have any doubts about the importance of IQ, read Gottfredson's "Why *g* Matters" or Robert Gordon's "Everyday Life as an Intelligence Test" in the special issue of *Intelligence*.

In the category of "What's to be Done?" the scientific answer agrees with Freedman and with Hunt's conclusion that we need to get at the hyphen, "to go beyond statistics to investigate the mechanisms of genetic action, including the interaction between genetic predisposition and the environment...to understand the causal pathway, both the molecular genetics and the psychological-social end of the continuum between genes and behavior" (p.549). At the level of public policy, the most important contribution in these five books is Detterman and Thompson's article in the *American Psychologist* which notes that "if the fit between the educational system and the learner is what is at fault, giving the learner greater exposure to the educational system is unlikely to produce substantial and lasting effects. The only way to do that is to build a system that suits the student's characteristics" (p.1087). But they also note that "there is currently a tremendous financial and emotional investment in business as usual. Schools, teachers, professionals, drug companies, and testing companies have a sizable investment in the status quo. It is easier to do things the way they have always been done" (p.1089). What financial incentive takes away, however, it also gives back. The multi-media computer technology now exists to provide lifelong educational resource for every citizen, tailor-made to that individual's strengths and designed to improve his or her weaknesses. Clearly the market niche exists, and parents with the resources will jump at the chance to provide its products for their children. And in the age of shrinking budgets, skyrocketing costs, global markets, and international economic competition, government needs to look carefully at the option of providing the resources for those who can't afford them. But those musings belong in a review of two other books by authors whose work has been mentioned here – Earl Hunt's *Will We Be Smart Enough?: A Cognitive Analysis of the Coming Work Force* (Russell Sage Foundation, 1995) and Richard Lynn's *Dysgenics: The Genetic Deterioration in Modern Populations* (Praeger, 1996).

Frank Miele

CHAPTER 13

DOUBLE VISION

A REVIEW OF:

ENTWINED LIVES: TWINS AND WHAT THEY TELL US ABOUT HUMAN BEHAVIOR

BY NANCY L. SEGAL. (NEW YORK, NY: DUTTON, 1999) 396 PP. HARDCOVER. $27.95. ISBN 0-525-94465-6.

Figure 13.0

In 1875 Charles Darwin's cousin, Sir Francis Galton, inaugurated the scientific study of human differences. He proposed that the two types of twins, identical and fraternal, provide a way to distinguish between "the effects of tendencies received at birth, and of those that were imposed by circumstances…in other words, between the effects of nature and nurture" (quoted on p. 2). Identical twins develop from a single fertilized egg that splits in two and are identical. Fraternal twins develop from two separate eggs fertilized by two separate sperm and are no more alike than ordinary siblings. By comparing the relative similarity among a number of identical twins, versus a group of fraternal twins, it is possible to get a rough gauge of the effects of heredity and environment. If identical twins are much more alike than fraternal, it suggests that there is an important genetic component in the characteristic being studied. One obvious problem in using this method with behavior is that because identical twins look much more alike, they may be treated more alike, and this could produce their greater behavioral similarity. For this reason, the most valuable research comes from looking at those rare cases of identical twins who have been separated early in life and reared in different home environments. The scientific research on such twins, their remarkably similarities in even minor aspects of their lives, and enlightening photos, are all in this volume.

Entwined Lives is far more than a book on what twin studies tell about the role of heredity in human behavior. As a former assistant director of the Minnesota Twin project, who has both conducted empirical studies and developed new methods of analysis, Nancy Segal is well qualified to provide a thorough and up to date review of what twin studies tell us. Independent studies conducted in a number of countries all demonstrate a genetic influence for general intelligence of 50-70%, 50% for the Big Five personality factors, 40% for occupational interests, and 34% for social attitudes (p. 314). Once popular anti-genetic criticisms, for example that Sir Cyril Burt had supposedly "cooked his numbers," no longer hold water. First, Segal points out, "the consistency of the IQ findings from the different identical reared apart twins is compelling because the information was gathered by different investigators, using different tests, in different countries, at different times, and from different twins" (p.136). Further, the accusations against Burt have themselves been scrutinized and "while some questions remain, two investigations published in 1989 and 1991 [Joynson, and Fletcher] cast doubt on Burt's supposed scientific transgressions" (p. 137). Finally, the Human Genome Project is providing direct evidence for the genetic component of human behavior. As Hillary Clinton, hardly an arch-hereditarian, noted: "If we pick up any magazine or newspaper these days, these are the headlines we're likely to find: 'Twins Unlocking the Secret of Identity.'"

Segal also leads her readers through the borderland emerging disciplines of evolutionary psychology, behavior genetics, and developmental genetics. The real questions in behavioral research today are not whether it's nature or nurture, but how nature works through nurture, and how nurture selects nature. If evolution selects for a specific set of adaptive behaviors, why is there so much behavioral variation, even within members of the same family? What role do within-family environmental differences, such as Sulloway's (1996) birth order effect, play? And how do all these factors interact during development? We still don't know the answers, but Segal has proposed a novel method to find out—"Studying identical twins reared apart who fall into similar and different birth order positions in their adoptive families" (p.68).

What makes this book especially valuable is that Segal tells her story from both the point of view of a leading researcher and of a twin herself. In addition to reviewing the latest findings on the genetic component in human behavior, other chapters deal with the special relationship between twins, bereavement at the loss of a twin, the effects of fertility treatments, famous sets of twins, twins in the animal kingdom, conjoined twins, mixed race twins, even sexual attraction between biological relatives separated early in life.

Figure 13.1: Reared apart identical twins, Jerry Levey (Left) and Mark Newman. The twins are reliving one of their first marvelous moments: comparison of physical characteristics soon after meeting. Note that the twins' little fingers suppor their cans of beer. Budweiser has always been their favorite brand. (Caption and photo from page 144 of the book.) Photo by Nancy Segal.

If there's a single message in *Entwined Lives* it is that each of us—even identical twins or conjoined twins—is a unique person. Knowing more about the genetic component in human nature can only help us best create environments that maximally nurture those abilities, allowing each person to develop to their fullest. Both behavioral scientists and general readers will enjoy and learn from this book.

References

Clinton, H. 1999. Introduction to White House Millennium Evening—Informatics Meets-Genomics (http://www.pub.whitehouse.gov/urires/I2R?urn:pdi://oma.e op.gov.us/1999/10/13/9.text.1)

Fletcher, R. 1991. *Science, Ideology and the Media: The Cyril Burt Scandal*. New Brunswick, NJ: Transaction.

Joynson, R. B. 1989. *The Burt Affair*. London: Routledge.

Sulloway, F. 1996. *Born to Rebel*. NY: Pantheon.

Frank Miele

Part IV

The Population, Environment, and Cloning Debates

Frank Miele

CHAPTER 14

SOULED OUT OR SOULED SHORT?

THE DEBATE BETWEEN ECOLOGISTS AND ECONOMISTS

MOTHER EARTH HAS BEEN SOULED OUT:

The present population of 5.5 billion is being maintained only through the exhaustion or dispersion of a one-time inheritance of natural capital....[including] high grade agricultural soils, ground water that accumulated during the last ice age, and biodiversity; each is being depleted globally at rates orders of magnitude in excess of regeneration and has no known substitute that could be feasibly supplied at levels close to those required....[I]t is evident that the human enterprise has not only exceeded its current carrying capacity, but is actually reducing future carrying capacity....Our descendants will have fewer of the essential requisites of life-support than we have today.
— Ecologist Paul Ehrlich

MANKIND IS BEING SOULED SHORT:

Environmental, resource, and population stresses are diminishing, and with the passage of time will have less influence than now upon the quality of human life on our planet.... Because of the increases in knowledge, the earth's 'carrying capacity' has by now no useful meaning. These trends strongly suggest a progressive improvement and enrichment of the earth's natural resource base, and of mankind's lot on earth.
There is only one important resource which has shown a trend of increasing scarcity rather than increasing abundance. That resource is the most important of all—human beings.
--- Economist Julian Simon

Before he became preoccupied making fund-raising phone calls from the vice president's office, then-Senator Albert Gore (1992) urged us to make the environment "the central organizing principle for civilization" (p. 270). In 1996, two books representing the adversarial positions on the question of human population and the carrying capacity of the Earth were published— *Betrayal of Science and Reason: How Anti-Environmental Rhetoric Threatens Our Future* by Paul and Anne Ehrlich, and *The Ultimate Resource 2* by Julian Simon. The two positions and their advocates differ in their methods, tone, and conclusion. SKEPTIC contacted both Ehrlich and Simon to ask them to be interviewed the debate on Population and Environment. Julian Simon accepted. (See Chapter 16) (Paul Ehrlich declined, though our offer remains open).

Economist Julian Simon's tome is 734 pages long, with 147 figures and 12 tables, and often reads like a prospectus for no-load, emerging-markets mutual fund. Ecologists Paul and Anne Ehrlich's book, which often reads like a combination of *J'accuse* and the Sixth Chapter of the *Book of Revelation* (i.e., the starting gate for the Four Horsemen of the Apocalypse), is 335 pages of text. My point is not the literary, analytical, or statistical skills of Simon or the Ehrlichs (which are all considerable), but rather the rhetorical styles in which the respective advocates have chosen to present their case before the jury.

In 1996 the paperback edition of Joel Cohen's *How Many People Can the Earth Support?* also appeared. While Cohen's perspective is much closer to Ehrlich's than to Simon's, his book provides an invaluable history and a wealth of data to be booked as evidence.

CHARTING POPULATION GROWTH

Figure 14.1 provides demographers' best guesstimates of the growth of human population from the end of the last ice age (about 10,000 years ago) to the present. (All figures, tables, and quotes are taken and adapted from Cohen, unless otherwise noted.) By that time, every continent save Antarctica was inhabited by only one kind of hominid—modern *Homo sapiens*. Our species had altered the existing ecosystems by introducing a new species (the dog) to Australia and North America and driven the megafauna of the Americas (giant sloths and camelids), and of Madagascar (lemurs the size of small bears) and possibly the Eurasian mammoths over the cliff of extinction, if not necessarily literally.

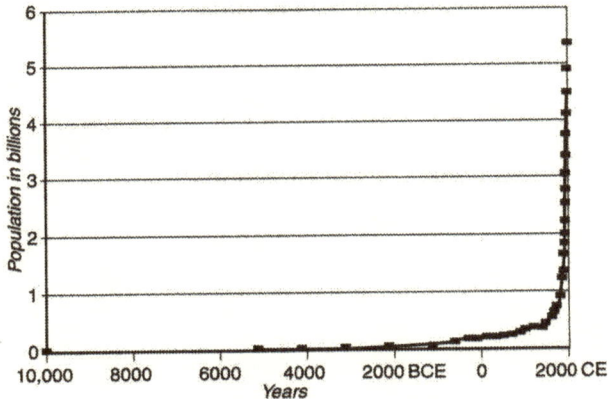

Figure 14.1: Estimated growth of human population over the last 10,000 years. (Adapted from Cohen, 1996; Figure 5.2, p. 81).

The graph shows the tremendous increase not only in the estimated absolute numbers of humans, but also the increase in the rate of increase. How accurate are such estimates? Cohen suggests within a factor of 10 at the start to a factor 1.05 at 1750 CE (Common Era). How can demographers estimate the population before the dawn of history? Primarily by multiplying the population density of known hunter-gatherer cultures in comparable climatic zones by the area known or believed to have been inhabited by ancestral populations. But the further back in time, the less certain the figures. One study contends that the relative lack of diversity in mitochondrial DNA implies that all living humans are survivors from a drop of the ancestral population from about 100,000 to only 10,000 people that took place around 80,000 to 100,000 years ago (Rogers and Harpending, 1993).

Figure 14.2 charts the historical increase in population from 1 CE to the present in greater detail. Note that these figures depict the estimates of total human population. The best estimates of different demographers vary, the rate of increase was not constant in all parts of the world, and even the overall graph does not describe perfect exponential growth.

Baron Rothschild is credited with having said, "I do not know the Seven Wonders of the World, but I have no doubt Compound Interest is the Eighth!" Table 14.1 at the bottom of the page shows the compounding of human numbers. What explains these increases in rate of growth and the concomitant increases in the absolute number of humans alive at any point in history? Obviously, the rate of birth and the rate of death. And what explains the differences in these rates for different societies and different periods in history?

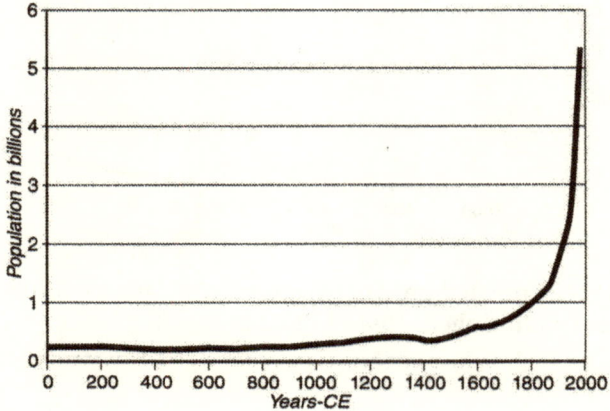

Figure 14.2: Esttimated growth of human population over the last 2,000 years. (Adapted from Cohen, 1996; Figure 5.3, p. 82).

Cohen explains the trajectory of human population (Figure 14.2) in terms of stages (Table 14-1), which he terms "evolutions," rather than "revolutions" to emphasize their gradual and haphazard, as opposed to abrupt and intentional, nature. The first three all produced increases in global population, the fourth produces declining growth because of reduced fertility. They are:

(1) The Local Agricultural Evolution, which took place at varying times from the end of the last ice age until 3000 BCE in Africa, Asia, Europe, Meso-America, and the Middle East;

(2) The Global Agricultural Evolution, which coincided with the Industrial Revolution and took place from 1650 to 1850 CE;

(3) The Public Health Evolution, which started around 1945 and is still taking place in some parts of the world;

(4) The Fertility Evolution, which began around 1785 in France and the United States and has yet to begin in some parts of the world.

Table 14.1- The Four Stages in Human Population Growth
(Adapted from Cohen, 1996; Table 2.2, p.30)

NAME OF EVOLUTION	TIME SPAN	POPULATION	DOUBLING TIME IN YEARS BEFORE	AFTER
Local Agriculture	10,000-3,000 BCE	0.005 billion	40,000-300,000	1,400-3,000
Global Agriculture	1650-1850 CE	0.75 billion	750-1,800	100-130
Public Health	1945-present	2.5 billion	87	36
Fertility	1785-present	3.7 billion	34 (peak)	more than 40

Figure 14.3 highlights the expected pattern in which the fall in birth rates lags behind the fall in death rates, and thereby generates an initial increase in population, followed by a later decline, as societies make the transition from Stage 1 to Stage 4. But this is by no means always the case.

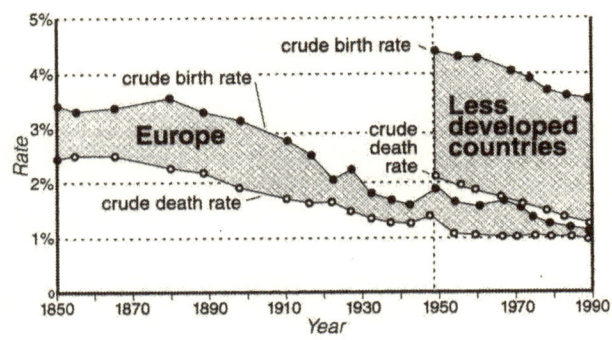

Figure 14.3: Birth and Death Rates since 1850. (Adapted from Cohen, 1996; Figure 4.4, p. 61).

Figure 14.4 shows how growth in world population reached its maximum rate of 2.1% per year between 1965 and 1970 and then began to fall, as more counties made the transition from Cohen's Stage 3 to Stage 4. This was primarily because of the drop of the growth rate in highly populated nations such as China, India, Indonesia, and Brazil. The absolute number of people, however, continued to increase, rising from 3.7 billion in 1970 to 5.3 billion in 1990. And even with the drop in the rate of growth, given the high absolute number of people, the number of souls added to the Earth each year rose from 72 million in 1970 to 92 million in 1990.

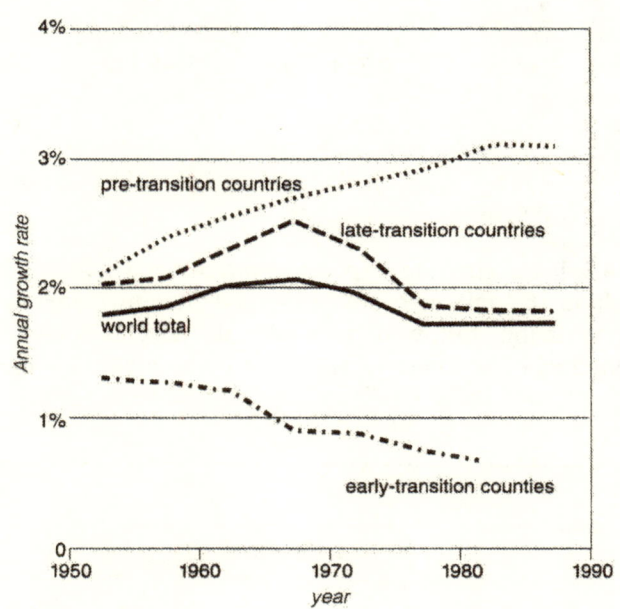

Figure 14.4: Global Trends in Population Growth from 1950 to 1990. (Adapted from Cohen, 1996; Figure 4.2, p. 55).

ESTIMATING CARRYING CAPACITY

Demographers use the term "carrying capacity" to designate how many people the Earth can support. If numeric population estimates are subject to error and disagreement, "guesstimating" how many people there ought to be reflects a serious conflict of visions. Cohen is again an invaluable source, summarizing the estimates, the methods used to derive them, and the limitations of the concept.

The concept of carrying capacity assumes that each additional person consumes an additional part of the Earth's resources and depletes (or at least alters) the environment, but also contributes through their productivity to the total resources available to others. Any reckoning of how many people the Earth can support depends upon how you estimate each term and the level of affluence at which you think the population can (or should) live.

Figure 14.5 shows the highest estimated carrying capacity presented by various scholars since Leeuwenhoek looked up from his microscope long enough to hazard a guess in 1679. Multiplying the surface area of the world by the population density of his native Holland at the time, he arrived at 13.4 billion. The subsequent numerical estimates range from Ehrlich's 1971 low ball figure of 0.5 to 1.2 billion, ("There are 3.6 billion human beings on the face of the Earth....between three and seven times more people than this planet can possibly maintain over a long period of time" [p. 410]) to Fremlin's intentionally-off-the-graph augmentio ad absurdum of 1018 (p. 408), where the only limitation is the ability of future technology to cool the heat generated by packing people as tightly as possible on the surface of the planet.

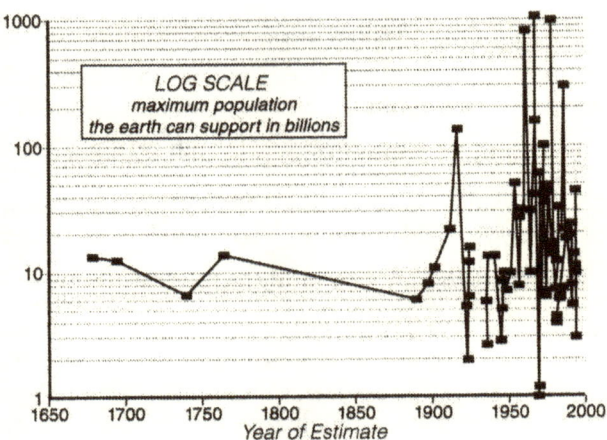

Figure 14.5: Estimates of Carrying Capacity. (Adapted from Cohen, 1996; Figure 11.2, p. 213).

IS CARRYING CAPACITY A MEANINGFUL CONCEPT?

According to Julian Simon and Herman Kahn the benefits of increasing human knowledge render the term "carrying capacity" devoid of "useful meaning" (p. 420). Garrett Hardin, who agrees with Simon on little else (see Chapter 15 and 16) also rejects the concept, arguing that "No thoughtful person is willing to assume that mere animal survival is acceptable when the animal is *Homo sapiens*." In its place, he offers the term "cultural carrying capacity," which includes supporting "the artifacts of human existence" and is therefore "inversely related to the quality of life" (p. 421). Lester Thurow amusingly draws attention to the degree to which carrying capacity is a function of human behavior and social organization:

> If the world's population had the productivity of the Swiss, the consumption habits of the Chinese, the egalitarian instincts of the Swedes, and the social discipline of the Japanese, then the planet could support many times its current population without privation for anyone. On the other hand, if the world's population had the productivity of Chad, the consumption habits of the United States, the inegalitarian instincts of India, and the social discipline of Argentina, then the planet could not support anywhere near its current numbers (p. 420).

Julian Simon and many economists look at the growth of human population and the accompanying overall increase in the standard of living and foresee the rise of humanity to an ever brighter future. For ecologists such as Paul Ehrlich and E. O. Wilson, and climate scientists like Steven Schneider, on the other hand, the future isn't what it used to be. They see an increasing population that is depleting the Earth's resources, irrevocably decreasing biodiversity, and driving, figuratively and literally, toward disastrous global warming.

What can account for the conflict between the economists' vision of the world as half-developed and the ecologists' vision of the world as half-destroyed? The answer lies less in the data we have reviewed than in what the disputing parties bring to the discussion table—the methodologies they employ, and the paradigms in which they work. We can see this more clearly by looking briefly at two counts in the ecologists' indictment—global warming and biodiversity.

GLOBAL WARMING

Steven Schneider is one of the leading climate scientists in the world. (He and Paul Ehrlich lost the famous wager to Julian Simon (Chapter 16 for the details and fallout.) Schneider's book, *Laboratory Earth*, provides an excellent summary of the evidence that supports the theory of industrially-induced climate change. Figure 14.6 (adapted from Schneider) shows the annual temperature (dots) over the last 150 years, shown as deviations using the 1951-1980 average as a baseline. There is a century-long warming trend (line) of about .5° C (.9° F), with the period 1981-1995 containing most of the warmest years on record.

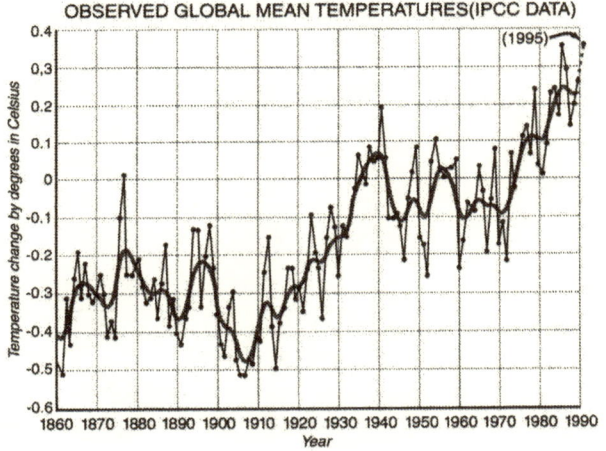

Figure 14.6: World Temperature Change 1861-1995. (Adapted from Schneider, 1997; Figure 3.1, p. 55).

Carbon dioxide (CO_2) absorbs infrared radiation, thereby trapping some of the Earth's radiant heat and raising temperatures (the greenhouse effect). How much of the century-long warming trend is due to ongoing temperature variation and how much is the result of the increased atmospheric CO_2 caused by the combustion of fossil fuels? Analysis of air bubbles trapped in ice cores from Antarctica shows that temperature and CO_2 levels have been highly correlated over the last 160,000 years, implying that the warming trend is not caused by human activity (see Figure 14.7, adapted from Schneider). But we also see the upward spike in temperature over the last 150 years, during which CO_2 has increased by 25%.

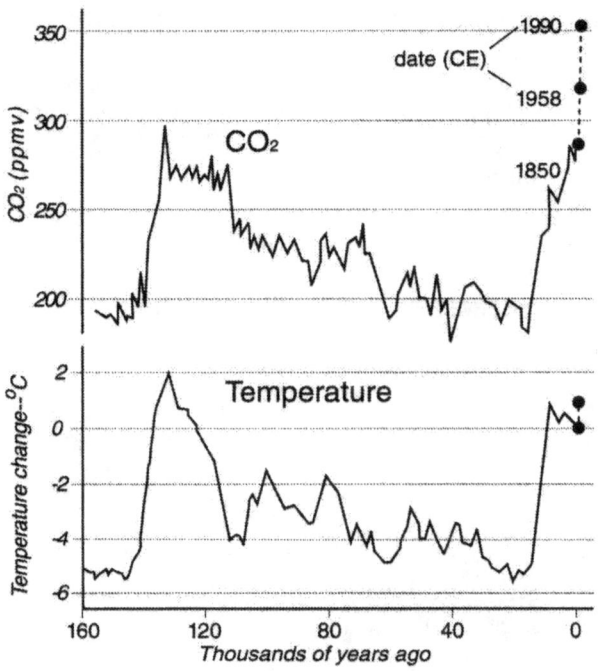

Figure 14.7: CO_2 and Temperature Over the Last 160,000 Years. (Adapted from Schneider, 1997; Figure 3.1, p. 48).

Schneider has developed a model to explain such climate change in terms of a positive feedback loop between CO2 and temperature. This implies that a small human-induced change could jolt the system into a more severe warming cycle. In 2050, when emissions have doubled the concentration of CO2 (or equivalent amounts of other greenhouse gases), the model predicts temperature change will occur at 10 to 50 times the ongoing rate, with potentially disastrous ecological and economic consequences.

BIODIVERSITY

The word biodiversity was coined by ecologists and biologists. They wanted to focus attention on the evidence that increasing human population and development, especially the clearing of the world's rain forests, were dangerously increasing the rate at which species were becoming extinct.

How many species are there in the world? What is the background rate at which they become extinct? How many have become extinct in the last 25 years? Nobody knows. Zoologist Robert May notes that science has a better estimate of how many atoms there are in the universe than of how many species exist on earth (May, 1992).

Biologist E. O. Wilson, the originator of sociobiology, has been the most eloquent advocate for the cause of preserving biodiversity. Wilson and Robert MacArthur developed the theory of island biogeography. They found that there was a consistent relationship between the size of an island and the number of species it contained, S = CAZ, where S is the number of species, A is the area of the island, C is a constant, and Z is a parameter whose value depends on the type of organism (bird, insect, plant, etc.) and the distance of the island from other islands (Figure 14.8, adapted from Wilson, 1992).

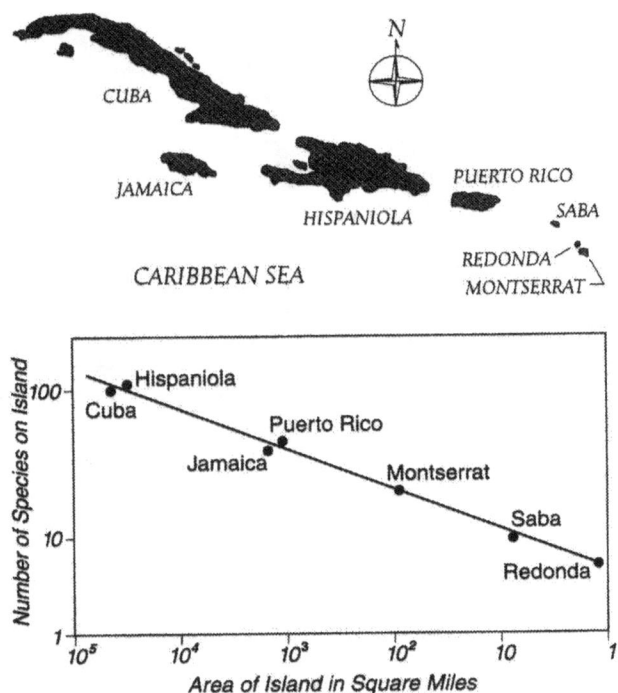

Figure 14.8: The Relationship between Area and Number of Species. (Adapted from Wilson, 1992; Figure on p. 222).

Wilson then tested these estimates by literally crawling over every square foot of some tiny islets of varying size, from the mud to the tree tops, and counting the number of species. He then had the island fumigated and studied recolonization. Other estimates come from comparing areas of Amazonian forest that have been converted into pasture against adjacent pristine areas of equal size. They show that when an area is reduced by 90%, the number of species drops by 50%. A greater number of extinctions probably takes place in rain forests, which are inhabited by a greater number of species than other areas.

Multiplying the amount of rain forest cleared as measured by satellite photographs and using modest values for the parameters, Wilson estimates that "human activity has increased extinction between 1,000 and 10,000 times" the background rate. He concludes that species are now going extinct at the rate of 27,000 species each year, or 74 per day, three per hour, and that we are "in the midst of one of the great extinction spasms of geological history" (p. 280).

"SHOW ME THE MONEY"

To arguments such as these, Julian Simon replies, "Show me the money!"— i.e., the shortages in commodities (as measured by an increase in the market price), the roster of species proven to be extinct, the famines brought on by global warming, or any of the disasters predicted by ecologists. His book is filled with graphs like Figure 14.9, which shows the fall in the price of copper (measured either against prices in general or hours of labor). Pick just about any commodity you want, Simon has a similarly shaped graph. If things are becoming scarcer, why aren't they becoming more expensive? Looking back on intellectual history since Malthus, economists sarcastically note that the end of the world has been coming for a long, long time.

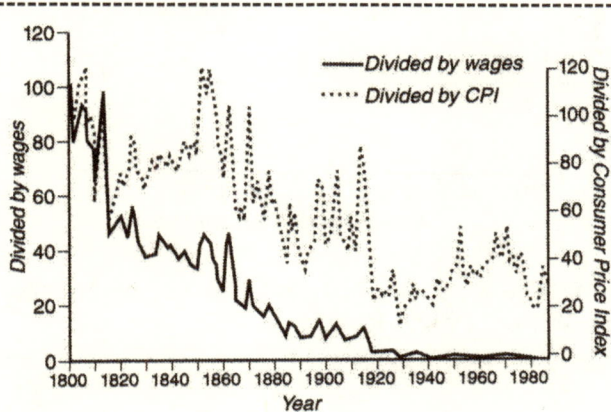

Figure 14.9: Price of Cooper Relative to Wages and to Other Commodities. (Adapted from Simon, 1996; Figure, 1-1, p. 25).

The reason we get such diametrically opposite closing arguments is that the opposing parties are asking different questions, using different methods to answer them, and basing their case on different underlying assumptions. Economists and ecologists differ not only in how they estimate impact, but in how they believe we adjust to it.

William Nordhaus is an economist (co-author with Paul Samuelson of the widely-used text, *Economics*) and a member of the National Academy of Sciences Carbon Dioxide Committee. He surveyed a group of experts (10 economists, four social scientists, five natural scientists, and two engineers) drawn from the NAS greenhouse-warming panels and asked them to estimate the likelihood of varying degrees of global warming and their economic consequences. He found that the natural scientists considered global warming much more likely, and that the adverse economic impact would be as much as 30 times greater than did the economists. One economist thought that the impact of global warming would be "small potatoes," while one of the natural scientists thought it would exceed the effects of the Great Depression.

When asked to explain such large differences, one of the respondents answered, "economists know little about the intricate web of natural ecosystems, whereas scientists know equally little about the incredible adaptability of human economies" (p.48). The ability of people, especially with modern technology, to find substitutes when the need arises lies at the heart of the economists' argument. While the price of copper has indeed gone down, the World Wide Web is increasingly being connected by even less expensive fiber optic cable, which also provides greater bandwidth and throughput. Perhaps the ultimate extension of the concept of adaptability is Simon's advice that even if thinning of the ozone layer were to increase UV radiation "people can intervene in many ways—even with as simple a device as wearing hats more frequently" (Myers and Simon, p. 63).

Steven Schneider tells his readers that the two most important questions to ask are simply: "What can happen? and What are the odds of it happening?" Julian Simon, on the other hand, told SKEPTIC, "We can never deal with the number of 'what ifs' or 'how abouts'." Schneider's climate forecasts and Wilson's extinction forecasts are based on extrapolation from the underlying theories of their respective sciences. To Simon and the economists, such underlying theories are irrelevant to forecasting human consequences, which are better estimated by past performance measured in terms of the market price.

EXPERT OPINION AND PUBLIC POLICY

Should we base our policies then on the demonstrated past history of market prices or on the risk avoidance models derived from scientific theory? Each has a problem.

If we take the market price as the measure of all things, at some point we must "internalize the externalities" so that the price paid reflects not just the costs of production and distribution, but also the cost of correcting any adverse environmental impact. The simplest mechanism is a tax. But at what rate? And who pays?

On the other hand, Ehrlich and Schneider both admit that their risk avoidance models do not produce the 5% level of confidence that is standard in statistical tests of scientific hypothesis. How do we weigh the obvious economic impact of regulation aimed at protecting the environment against potential downside risk predicted by the various models?

Since we have no idea how many species actually exist, any estimate of human impact beyond "a lot" is difficult to support. If no one knows how many species there really are, how can you know if one has gone extinct? How do you prove a negative? How do you know someone won't fish up a coelacanth? Increasingly ecologists have focused on variation within species and on ecosystems rather than species. How do you quantify those in a way that will satisfy critics?

What is certain is that the debate on whether the Earth has been souled out or humanity souled short is no isolated esoteric dispute. One way or another, it encompasses all of the following sub-debates: Abortion, Animal Rights, Endangered Species, Energy Alternatives, Environmental Justice, Euthanasia, Family Planning, Global Resources, Global Warming, Hunger, Immigration, Male/Female Sex Roles, Pollution, Poverty, Science and Religion, Third World Debt, and Trade.

Al Gore was half-right—the environment will become the central debating principle for the post-Cold War world. For most of history, traditional religion defined the relationship between humanity and nature. It remains to be seen whether the ecologists' vision or the economists' vision or some yet-to-be-worked out synthesis will become the central organizing principle for civilization in the next millennium.

References

Cohen, J. 1996. *How Many People Can the Earth Support?* NY: Norton.

Ehrlich, P. and Ehrlich, A. 1996. *Betrayal of Science and Reason—How Anti-Environmental Rhetoric Threatens Our Future.* Washington, D.C.: Island Press.

Gore, A.1992. *Earth in the Balance—Ecology and the Human Spirit.* Boston: Houghton Mifflin.

May, R. 1992. "How Many Species Inhabit the Earth?" *Scientific American.* Vol. 267. No. 4. pp. 42-49.

Miele, F. 1996. "Living Within Limits and Limits On Our Living. Garrett Hardin on Ecology, Economy, and Ethics." *Skeptic* Vol. 4. No. 2 pp. 42-46.

Myers, N. and Simon, J. 1994. *Scarcity or Abundance? A Debate on the Environment.* New York: Norton.

Nordhaus, W. 1994. "Expert Opinion on Climatic Change," *American Scientist. Vol.* 82, pp. 45-51.

Rogers, A. and Harpending, H. 1993. "Population Growth Makes Waves in the Distribution of Pairwise Genetic Distribution." *Molecular Biology and Evolution.* Vol. 9, No. 1, May.

Schneider, S. 1997. *Laboratory Earth—The Planetary Gamble We Can't Afford to Lose.* NY: Basic Books.

Simon, J. 1996. *The Ultimate Resource 2.* Princeton, NJ: Princeton University Press.

Wilson, E. O., 1992. *The Diversity of Life.* Cambridge, MA: Harvard University Press.

Frank Miele

CHAPTER 15

LIVING WITHIN LIMITS & LIMITS ON LIVING

GARRETT HARDIN ON ECOLOGY, ECONOMY, AND ETHICS

Garrett Hardin is a pioneer in the science of human ecology. He is best known for his 1968 article in *Science*, "The Tragedy of the Commons." Reprinted in over 100 anthologies, it is still cited by proponents of the free market as a classic analysis of the inherent failings in terms of economic loss and environmental degradation of public, as opposed to private, ownership. Cultural Survival, a quarterly journal that focuses on the rights of indigenous peoples and ethnic minorities, devoted an entire recent issue (Spring, 1996) to attempting to disprove Hardin's logic, arguing that, "effective use of common resources by local communities is in our day often the most efficient way of ensuring that modern, industrialized economies promote growth with equity and minimal environmental degradation" (Maybury-Lewis, p. 3).

But don't jump to the conclusion that Garrett Hardin is a patron saint of contemporary conservatives. To the pro-life, pro-growth movement, Hardin, a founding member of Planned Parenthood, early crusader for abortion rights, advocate for legalized euthanasia and assisted suicide, restricting immigration, reducing the world's population to 100 million and opting for sustainability rather than growth, seems almost Satanic.

For Hardin, our continued survival requires abandoning ideas of both equitable redistribution and laissez-faire in favor of living within limits. This, in turn, requires placing limits on living through what he straightforwardly terms "mutual coercion, mutually agreed upon." His recent book, *Living Within Limits: Ecology, Economics, and Population Taboos*, received the 1993 Phi Beta Kappa Award in Science. He is currently near completion of his twentieth book, *Lifeboat*.

While this is an election year, Hardin, who has no use for either the political right or the political left, certainly isn't running for any office. Rather, since retiring from the University of Santa Barbara in 1978, he has devoted his time to "stalking wild taboos" (as he titled one of his books, currently being revised for a third edition). Let us then pick up Hardin's trail and see where it leads.

Figure 15.0: Garrett Hardin

Miele: The word "ecologist" summons up an image of a caring person concerned about the whole planet, if perhaps somewhat unrealistically. By contrast, the dismal science of economics seems tough-minded, self-centered, and deadly practical. It's hard to imagine someone who describes him or herself as "ecologically-minded" answering a personal ad from someone who is "economically-minded." What is the relationship between ecology and economy?

Hardin: The view that I and a number of other ecologists share is that ecology is the overall science of which economics is a minor specialty.

Miele: Do economists accept that definition?

Hardin: No, they don't. I was just talking with an economist friend of mine about the bet that population ecologist Paul Ehrlich lost to economist Julian Simon. Simon wagered that the price of certain commodities would be no greater five years from the time of the bet and maybe would be less. Ehrlich should never have made that bet because it accepts the economists' definition that the true price of something is the published market price. The ecological definition states that the price of producing something is not only that market price, but also the price of all the things you have destroyed in the process of bringing the item to the market.

Think of the state of West Virginia. Three hundred years ago it must have been a paradise. It's been made into an absolute shambles by mining. Mining generally has been an extremely destructive activity. The mining interests by and large pay few if any of the costs of what they've destroyed. In fact, they often get subsidies. So the market price of mined commodities is an utter fiction. The true ecological price, which would be very difficult to determine, would reflect what we are actually paying to dig all these things out of the ground.

In general, economic analysis is poorly fitted to deal with the future. The method of discounting in effect discounts the distant future to zero. Well, if there's one thing we know, it's that there will be a distant future. By training, economists are not equipped to deal with the future and they are not equipped to deal with the total system. They just deal with one human and then another and they constantly get in trouble when they forget about the environment.

But one shouldn't lump all economists together. I could list at least a dozen economists who have what I call an ecological point of view. One of the most prominent is Herman Daly of the University of Maryland.

Miele: Then he's right near Julian Simon.

Hardin: Geographically he's near Julian Simon, but only geographically.

Miele: But don't economists take account of the environment under the concept of externalities?

Hardin: Oh, absolutely. But externalities are things they just don't want to see. Once they invoke the term "externalities," they have sprinkled holy water on the problem and in their calculations it no longer exists and therefore has no effect in our decision-making processes.

Miele: Don't economists also use the concept of opportunity costs, that is, the offsetting cost of the lost opportunities of not doing something?

Hardin: I think that concept should be dismissed almost entirely because you can bring in anything through the escape hatch of opportunity costs.

Miele: You mentioned the effect of subsidies. We're here in Southern California, not far from one of the most currently productive agricultural areas on Earth. Is there anything ecologically or economically more nonsensical than using subsidized water to grow rice and cotton on what would otherwise be a desert?

Hardin: Yes. Subsidizing the export price of the cotton, which is what we do!

Miele: What then is the relationship between ecology, economy, and ethics?

Hardin: Ethics gets involved because we are not the Man from Mars. The Man from Mars could study the ecology and the economics of all our processes on Earth and never give a damn what answer he got. It's just not his problem. But we're not the Man from Mars. We are living here on this planet and so we do give a damn. If we make a mess of the Earth, we still have to live on it. If I have one idea of the best life and someone else has a different idea and we then try and reconcile the two, we enter the realm of ethics. I would define an ethical question as one in which we have a choice between alternatives for which it is hard to find a common measure, but among which we must still choose.

Biological ethics is more than just ethics applied to biological problems. Sometimes it is called "toughlove ethics," built on a biological foundation. Its essential elements are relative quantities, feedback processes, and the changes that time brings forth as unforeseen consequences of our actions. The tools required for biological ethical analysis are literacy (the correct use of words), numeracy (dealing with quantities) and ecolacy (the study of relationships over time).

Miele: Suppose that Man from Mars was intelligent and very long lived and looking through a very powerful telescope. Might he get the impression that since the dawn of industrial, market-based economies, humans, especially Europeans, for the most part, have spread across the Earth like a cancer? If he could do some sophisticated analyses of his observations, he might find that one segment of one species was using more and more of the biosphere's resources and converting them into more of his own kind. Would the Man on Mars be far off the mark?

Hardin: He would be right that the human species viewed as a whole has been a disaster for the Earth. Innumerable deserts have been made primarily by man's activities. The weather contributes, but the two work together. If you look at pictures or read accounts of Arizona and New Mexico from 100 years ago, you'll discover there wasn't anywhere near the level of desertification that you can see there today. And the desertification is a result of our activities. Industrial, market-based economies are motivated to get profits, where a genuine subsistence economy will not go that far. A genuine subsistence economy will not dig a deep well. It just doesn't pay. They dig shallow wells, but not deep ones.

But subsistence economies can crash if they take out of the environment faster than the environment can repair itself. Desertifications have happened time and again as the result of having too many plant-eating animals. North Africa is a memorial to the goat. In Roman times, it was the granary of the empire, now it's a desert. On Easter Island, they just had too many people and they didn't see the consequences of cutting down all the trees.

Miele: But doesn't evolutionary theory tells us that every species, every individual, indeed, every gene tries to turn the resources of the environment into replicas of itself? Isn't that why we're here?

Hardin: That's right. But our success may kill us. We have brains that are capable of doing fantastic things no other species can. We have developed technology that speeds up the process. But as a group we haven't accepted the results of certain observations. The economists and the technologists have become possessed by the Myth of Perpetual Growth and they think it can go on forever. It is quite possible for our species to commit suicide.

Miele: I recently read a book on the Rwanda genocide. After describing the history, politics and bloody details of the crisis, the author stated (Prunier, 1995, p. 353):

> There is of course one further added cause: overpopulation. This is still a taboo, because human beings are not supposed to be rats in a laboratory cage and Christians, Marxists, Islamic fundamentalists and World Bank experts will all tell you that overpopulation is relative and that God (or modern technology of the Shari'a) will provide. But let whoever has not at least once felt murderous in a crowded subway at rush-hour throw the first stone.

You once wrote a sarcastic piece entitled, "Nobody Ever Dies from Overpopulation." Would you care to comment?

Hardin: It's absolutely true. There should be a formal scientific investigation by people who are trained in that sort of thing, which I am not, of the change in attitude toward the subject of over-population and how it came about. There was a lot of concern about population in the first part of this century. It culminated, as I see it, in Paul Ehrlich's book, *The Population Bomb*, in 1967. Then all the attacks started from various economists and the public just got tired of it as they became aware of the fact that the people who were talking about over-population were not offering any remedies. The feeling on the part of the public was "What good is it seeing a problem, if you don't offer a remedy?" The same criticism has been made of my last book, *Living Within Limits*. Right now we've reached a low point in getting people to take population problems seriously. But I have been encouraged by the letters to the editor I see in the local papers. The best summary of the problem is the one Ehrlich uses—name any problem that you regard as important and dangerous and you will find that unless you solve the population problem, you can't solve that one either. Population is not the sole cause, but it's a contributing cause to all the other major problems.

Miele: In the early sixties, former presidents Harry Truman and Dwight Eisenhower were the honorary co-chairmen of Planned Parenthood. Planned Parenthood clinics are now being routinely picketed and in some cases bombed. How has family planning become a dirty word?

Hardin: I started being an activist for legalized abortion in 1963. I spent most of my external time on that issue until the Supreme Court reached the famous 1973 *Roe v. Wade* decision. I thought the fight was all over. Well, I was wrong. At the time, my wife and I were active in Planned Parenthood. She was on the local board of directors. The question came up in Planned Parenthood as to what our position should be on abortion. Some wanted to stay clear of it entirely because they realized there would be a lot of opposition. Fortunately, Planned Parenthood decided that it was a question of women's rights. Abortion is above all other things a method of birth control. To put it another way, it's a backstop for any system of birth control when the rest of the system fails. That decision to support the woman's right to abortion put Planned Parenthood in a dangerous position. As opposition developed, the opponents then went on to say that everything Planned Parenthood was doing was just window-dressing for their principal interest—killing babies. No president could now accept such a position.

Miele: How do you answer the objection of the anti-abortionists that abortion puts us on the slippery slope to euthanasia, assisted suicide, and elimination of the "unfit?"

Hardin: First let's look at the concept of the slippery slope. Every ethical decision puts you on the slippery slope. You just have to live with it. For example, we used to have a speed limit of 65 mph. That's a slippery slope for God's sake. No matter where you put the speed limit, people want to push it up and up and up. The only completely safe speed limit is zero mph. Anything above that will get pushed up and up. What do we do? We draw an arbitrary line and set the speed limit at a certain level. Back in the days of the old 65-mph limit, you couldn't prove statistically that 66 was more dangerous than 65.

In the specific case of abortion, the matter is particularly easy in that no woman wants a late abortion. Once abortion was made legal, the age of the aborted fetus went down. The slope slipped in the other direction. If we legalize RU-486 and other similar new drugs, the age will fall to one week or less and start approaching zero. The slippery slope will slide in the other direction. The only reason we have late abortions is because we make early abortion difficult.

Miele: What about the issue of President Clinton's veto of the bill banning what are called "partial birth abortions?"

Hardin: This is a very special case. Almost no attention should be paid to it. You don't have to pass a law about this, for God's sake. The medical profession only performs this procedure in cases where the alternatives are much worse. But then slippery slope gets invoked to argue that this will lead to infanticide.

Miele: Let's consider the case of an anencephalic fetus or child (that is, one in which the brain fails to develop).

Hardin: The only question we should ask is "What does the woman want?" If she wants an abortion, do we have a good reason for denying her that? I don't think we ever do. If the woman doesn't want to abort an anencephalic fetus, that's a tragedy. But she will have to live with that decision. Society should be very chary about taking a hand in forcing a woman to have an abortion in such a tragic case.

Miele: But what about euthanasia of an anencephalic child? Should that be allowed?

Hardin: Yes. The point we must realize is that the bringing up of children was once a function only of the family. It is now a function of the entire community. And that can be put in economic terms. Bringing up a normal, healthy child to age 16, without any college education, costs $100,000. The parents pay some of that cost, but the community—all taxpayers—pays part of that cost. We allow the parent to decide to have the child, but the parent doesn't pay all the costs.

Miele: Do you think that gives society a right to speak on this issue?

Hardin: Since society pays the costs, yes. In the case of an anencephalic child, the parents can't possibly pay the costs. Society pays the cost and society can't possibly have any reason for wanting an anencephalic child.

Miele: Well, there are religious reasons.

Hardin: Religious reasons, which is no reason. I notice *Skeptic* had a review of Dennett's book, *Darwin's Dangerous Idea*. Religious reasons amount to what Dennett terms "skyhooks." Do you believe in skyhooks? I don't.

Miele: What about assisted suicide, now best known because of Dr. Kevorkian's legal battles?

Hardin: Look, I'm 81. I may be wanting that one of these days. Because of polio, I'm dependent on my arms. I can't walk. If I lose my arms, should I get in a wheelchair and have to have somebody push me around all the time? At that point, I'm going to be looking for Dr. Kevorkian. I don't want to sit in a wheelchair for the rest of my life.

Miele: But your mind is still sharp and you have contributions to make.

Hardin: Ah, thank you for those kind words. I know how un-sharp my mind is becoming. And though I appreciate the tact of other people, looking at my fellow oldsters, I know damned well we're going downhill. So, by association I say, "probably, me too."

Miele: Reading your books or Paul Ehrlich's, one gets the feeling we're going to hell in a hurry rather than a hand basket. But doesn't Julian Simon have a point that since Plato, and certainly since Malthus, you doomsayers have always said this and you've always been proved wrong?

Hardin: Not always proved wrong. Some of the things that were said have proved right.

Miele: For example?

Hardin: Here we get into an argument. Many people think that living on Manhattan all the time is not the ideal way to live. Society paints itself into a cul-de-sac of urbanization. I grew up on a farm in Missouri. When I was very young, I remember the creek running all summer long. By my teens, it ran only at the beginning of the summer and left a hole for the rest of the summer. By the mid-30s, even the water-hole was gone. My grandfather told me that when he first moved there, the water was so clear and you could see to the bottom. As far as I was concerned, all water, in creeks, rivers, and lakes, was brown. So it was a shock when I went to England and Ireland in my 50s and saw small rivers that were so clear you could see the fish.

Miele: But the argument against your story is that we now have, along with that somewhat polluted water, more people, more goods, more services. More people are living better than ever before. So what's one dirty pond versus 100 more people with good jobs raising their families?

Hardin: The important thing, and on which you can't get agreement, is the question, "Why is having more people good?" Now, if you say more people produce more goods so that there is a higher standard of living, then you've got a good argument. But suppose more people is just more people. Suppose more people means a lower quality of life? Do we simply want to maximize the number of people?

Miele: From a Darwinian perspective, more people allows a more stratified society and more specialization of labor. In the long run, those more populous, stratified societies will displace the less populous, more egalitarian ones. It's going to be that way, whether we like it or not.

Hardin: You may be right and this is the tragedy. Can we stop our built-in suicidal tendencies? All species have the blessing of enemies that keep their numbers under control. But we have been getting rid of all our enemies. We thought we'd gotten rid of all disease. We haven't quite, but if we can, by God we will. Having gotten rid of all the lions and tigers and bears, when we get rid of the micropredators, the bacteria, and the viruses, there will be nothing to control our numbers except ourselves. If not, we will commit suicide. The Easter Island tragedy will be universalized. There will be a few people, leading a miserable life.

   If we control our numbers, we might be able to settle on a world population of up to 100 million, living a hell of a good life. But we're up to six billion now. That's six thousand million. This level is impossible to maintain.

Miele: There is a certain self-satisfied negativity to much ecological writing. Julian Simon has written other books about starting your own mail order business and how to overcome depression. I just can't imagine either you or Ehrlich writing anything so upbeat and optimistic. Isn't that an inherent flaw in the ecological movement? Aren't you going to lose the argument by induced depression as your readers reach for the Prozac?

Hardin: If your only goal is to win, the answer is to become another Julian Simon. I couldn't live with myself if I did that. I'd die of shame. While I may be a bearer of bad news, I try to convince people that what sounds like bad news is better than "good news" that's wrong.

Miele: Suppose you're 100% correct in everything you've said. Those who listen to you will control their family size. But those who don't, won't. So again the pro-growth side wins by default.

Hardin: That's right. I resigned from Zero Population Growth because I realized that the only people paying attention to ZPG were college girls who decided they'd have no babies. So ZPG was self-extinguishing. The only answer is that family size cannot be left to individual decisions. You don't have to be brutal about it. You can use incentives. But control of population will have to take the form of mutual coercion, mutually agreed upon.

Miele: A good argument can be made that "ecologically-minded global consciousness" is a luxury of upscale individuals in first world countries and that its principal target is third world populations. Having already raped and plundered a good part of the planet, according to your own and similar accounts, who are we to lecture the rest of the world on anything?

Hardin: Lecturing is a difficult art. You have to get people willing to be lectured to. My basic position is, never globalize a problem if it can possibly be dealt with on a local basis. We could form a global committee to fill in the potholes of the world and the result would be a disaster. Imagine if you had a pothole in front of your house and you couldn't get it filled in until you got approval from some council thousands of miles away representing six billion people. Don't be silly. Fill in the pothole yourself.

Miele: Isn't that the problem with the United Nations?

Hardin: Yes. We have to localize problems, not globalize them, and reproduction is a very localized problem. It takes place in a very local place every time—in a bed. Some things, like the destruction of the ozone layer, are a worldwide problem and I don't know what the answer is.

Miele: Do you think that the ozone hole is a realistic problem?

Hardin: If it isn't now, it will be by the time we have 12 billion people. If it isn't at 12 billion, it will be when we have 50 billion people. The point is that we are doing all sorts of things to our common air and water, to the oceans and the atmosphere. And at some point it will do us in unless we find ways of mutually coercing each nation on these matters.

Miele: Do you consider global warming a serious danger? I've seen accounts on both sides and they look like computer simulations that give you whatever you want.

Hardin: The information is just at the edge of what you can say is thoroughly reliable. Since we don't really know, the question is what do you do when good arguments can be presented either way? Prudence would dictate that you take the pessimistic answer. The reason economists don't want to do that is because they see anything that cuts down on economic activity as an evil, and with some justice. But I think we have to accept that evil, rather than a worse one.

Miele: Haven't Repetto and others made a convincing argument that the only way to slow down the rate of population growth in third world countries is by making them richer, not poorer, and better educated, rather than less educated, especially for females?

Hardin: The question is how to lower the growth rate in the third world. First of all, we have to get rid of the argument of the benign demographic transition that held sway from about 1935 to 1975. It was based on the fact that the period during which Europe became more prosperous was also the period in which Europe lowered its birth rate. From that it was deduced that if you become more prosperous, you will lower your birth rate. If that were true, all you would have to do is shovel money into a poor country and their birth rate would go down. Well, we've been doing our best, shoveling money into poor countries and the result has been the opposite. Maybe their birth rate goes down, but the death rate goes down even faster, so that their net survival rate goes up and they end up worse off than they were.

Miele: What about the effects of female education?

Hardin: Yes, there's good evidence for that. If we can have a hand in increasing other occupational opportunities for women, we should, because experience shows that it works.

Miele: Your recent book is on the controversial issue of immigration policy. What is your position?

Hardin: Admitting immigrants from over-populated countries amounts to taking on their problems which they haven't solved. If we take on their problems, they will never solve them by other means.

Miele: An anthropologist named Virginia Abernethy has recommended weighting the immigration quotas for various nations inversely proportional to their fertility rate.

Hardin: I think that's a very good idea. At least we wouldn't be harming those other nations by encouraging them to over-populate.

Miele: Can the world reach a balance between sustainable economic growth and environmental degradation?

Hardin: The dangerous concept of sustainable growth was introduced into the debate about eight years ago. Sustainable growth is an oxymoron. Don't go for growth, go for a steady state, a term that comes from biological study. There's always an increase, but there's also a decrease around a mean value.

Miele: But doesn't that mean condemning a large portion of the world's people to a life of misery with little hope?

Hardin: They're condemned in either case. Sustained growth condemns them to that too. In America, we were able to expand into a vacuum, but that's at an end now. And certainly for the poor countries, they're ain't no vacuum for them to expand into.

Miele: But we come back to the question of how you can sell that to anyone?

Hardin: Maybe I can't. Maybe all I can do is ask, "Which do you prefer, early suicide or late suicide? You have the choice." We have to get the world to accept the idea of limits. It goes into every decision we make. This will not be anything new. It will be a return to the conservatism of humanity for all but the last 500 years. We can have change at the intellectual level, where the resources involved are ideas. But there are limits on material resources. And we have to learn to live within limits. From an evolutionary perspective it amounts to realizing that you cannot escape natural selection.

And what is natural selection? It is the natural investment in success. It is much broader and more revolutionary than just biological science.

Miele: Thank you for taking the time to speak so directly with *Skeptic*.

References

Abernethy, V. 1993. *Population Politics.* New York: Insight Books.

Baden, J. and Hardin, G. 1977. *Managing the Commons.* San Francisco, CA: Freeman.

Beck, R. 1996. *The Case Against Immigration.* N.Y.: Norton.

Daly, H. 1977. *Steady-State Economics.* San Francisco, CA: Freeman.

Ehrlich, P. 1968. *The Population Bomb.* New York: Ballantine.

Hardin, G. 1968. "The Tragedy of the Commons," *Science,* 162:1243-1248.

____. 1993. *Living Within Limits—Ecology, Economics, and Population Taboos.* New York: Oxford.

____.1995. *The Immigration Dilemma: Avoiding the Tragedy of the Commons.* Washington, D.C.: Federation for American Immigration Reform.

____. 1996. *Stalking the Wild Taboo.* (3rd edition) Petoskey, MI: Social Contract Press, in press.

Maybury-Lewis, D. 1996. "Voices from the Commons—Evolving Relations of Property and Management." *Cultural Survival,* Spring, p. 3.

Prunier, G. 1995. *The Rwanda Crisis—History of a Genocide.* New York: Columbia University Press.

Repetto, R. 1979. *Economic Equality and Fertility in Developing Countries.* Baltimore: Johns Hopkins University Press.

Simon, J. 1981. *The Ultimate Resource.* Princeton, N.J.: Princeton University Press.

____. 1989. *The Economic Consequences of Immigration.* Oxford: Basil Blackwell.

CHAPTER 16

LIVING WITHOUT LIMITS
AN INTERVIEW WITH JULIAN SIMON

AN ECONOMIST EXPLAINS WHAT'S WRONG WITH THE ENVIRONMENTAL DOOMSAYERS, THE UNION OF CONCERNED SCIENTISTS ...AND SKEPTIC MAGAZINE!

Julian Simon, Professor of Business Administration at the University of Maryland and Senior Fellow at the Cato Institute (a libertarian think tank), has been described as his day's "leading anti-neo-Malthusian." In his books, *Population Matters: People, Resources, Environment, Immigration*; *The State of Humanity*; and *The Ultimate Resource 2*, he has argued that demographic, economic, and statistical analyses show doomsday predictions that overpopulation has taken the Earth to the brink of ecological destruction, environmental pollution, mass starvation, or species extinction are without any basis in fact.

In 1990, Simon won his famous wager with ecologist Paul Ehrlich that supplies of five metals, as measured by a decrease in their market price, would increase in the ten-year period 1980-1990. Writing about the wager in the *San Francisco Chronicle* Simon stated that, "Every measure of material and environmental welfare in the United States and in the world has improved rather than deteriorated. All long-run trends point in exactly the opposite direction from the projections of the doomsayers." (The so-far unaccepted offers by both Ehrlich and Simon to "go double or nothing" are described in the interview.)

Attempts to calculate the Earth's "carrying capacity," or the "total reserves" of any material are, in Simon's view, as inherently theological and meaningless as attempts to determine how many angels can dance on the head of a pin. An advocate of both increased population and increased immigration to the U.S., Simon insists that economic analysis proves that the only resource that has shown a trend toward increasing scarcity is the most important resource of all—people, all people. Since more people also means more human ingenuity, he contends, information and technology can provide for infinite growth.

Figure 16.0: Julian Simon

Miele: Our magazine is dedicated to being skeptical—of mainstream expert opinion as much as either fringe opinion or "common sense." The issue in which this Q&A will appear will contain the World's Scientists' Warning to Humanity. You are familiar with its content. Why should we be skeptical of the Warning?

Simon: Given the name and spirit of the magazine—*Skeptic*—it seems to me that the first task of a good skeptic is to make good judgments about which evidence is sound and which evidence is not sound, and the next is to decide whose statements to trust and whose not to trust. We prefer not to just "trust" other people, but you can't check out everything directly.

The Union of Concerned Scientists' Statement and the statement by the National Academy of Science (NAS) are flawed on both counts. First of all, neither of these statements contain data, but instead contain opinions. Even worse, these statements are not produced by people who are scholars of the subject. They are instead filled with signatures of high-powered Nobel laureates in all kinds of other fields. But they are not statements by population economists.

It is very important to recognize that we're talking about the economic consequences of things, which is the particular expertise of population economists. It is not just the absence of population economists that should make one skeptical of such statements, but the fact that among population economists, if not unanimity, there is at least a very strong consensus in the direction opposite to the statements. The NAS Statement itself is utterly the opposite of the NAS Report, which was produced by scholars of the subject in 1986. That report, by the way, is referred to by the NAS executives as a "problem report," which means it didn't do what they expected it to do. In my book, *Population Matters*, I document how they rigged the press release and twisted the arm of the main author of the report, Sam Preston, the night before the press conference to make what he had to say more radical. The NAS executives later produced a statement, not a research report, by all these other people who are nonpopulation economists, that presents all the usual scare stuff.

So my question to you is, why is SKEPTIC publishing such a report if it fails these tests of good evidence?

Miele: On issues of scientific controversy, we like to present a range of opinion by authorities in the field. That's why we wanted to get your evaluation of the statement.

Simon: Yes, but the stuff in that statement is not by authorities. That's precisely my point.

Miele: Well, that statement was signed by over 1,670 scientists, including 104 Nobel laureates, many in the disciplines that are directly concerned with the study of biological phenomena, which you are not. So why should we believe economist Julian Simon, Professor of Business Administration, about matters of biology and the environment, rather than biologists and ecologists such as Paul Ehrlich and E. O. Wilson, who are the leading scholars in those fields?

Simon: I'm not a biologist, that's true. But the subject of these reports is not biology, but rather economic human consequences—what will be the effect of population on the food supply, what will be the effect of population on the supply of natural resources, what will be the effect of population as to whether our air and water will get cleaner, rather than dirtier. Those questions have nothing whatsoever to do with biology, but instead depend on two branches of thought. One is economics; the other is statistics.

Miele: What about the question of biodiversity? Isn't that a biological question?

Simon: It depends what the question is. If the question is, "How many species are going extinct?" That is a statistical matter, not a biological matter.

Miele: But consider the question, "What makes up an ecosystem and what role do each of those different species play in that system?" Would you call that an economic question or a biological question?

Simon: I would assume that is a biological question. But that is not the subject of these statements. The subject of these statements is that resources are disappearing and becoming more scarce and that our air and water are becoming dirtier. Let me summarize the relevant findings. Every agricultural economist in the United States and probably most of the world knows that the world has been eating better, decade by decade, since World War II.

Miele: How do you define "eating better"? Are you talking about gross numbers or percentages? The gross number of people eating well is probably going up, but so perhaps is the number of people starving, simply because we have so many more people than we did 50 years ago.

Simon: We can never deal with the number of "what ifs" or "how abouts." The only way to answer the question scientifically is for you as the skeptic to provide a set of sample objections and I as the population economist will then provide the relevant data. If all the data contradict the sample objections, then you, as skeptics, have got to conclude the statement is in error.

The example you just provided—famine deaths—is an excellent starting point. It has been studied by a very good agricultural economist, D. Gale Johnson at the University of Chicago. He probably is second only to his teacher, Theodore Schultz, who won the Nobel Prize in 1979. Johnson writes in *The Encyclopedia Britannica* that even though there are many, many more people alive right now than there were in the last quarter of the 19th century, perhaps 10 times as many people—not percentage, but absolute numbers—died from famine in the last quarter of the 19th century.

Next, we know from the studies of agricultural economists that the average number of calories per person is going up all over the world. Even more important, we know that people are living longer, which is very related to their nutrition.

Miele: Is that simply because of a decline in the infant mortality rate, which would increase the mean life expectancy?

Simon: It could be, but it's not. People are living longer at every age. This is the domain of demographers, not of biologists. I'm a demographer, a statistician, and an economist. Yet biologists speak about these matters even though it's is not their trade and many of them don't understand demography. Every demographer knows that people have been living longer, decade by decade. In the rich countries life expectancy has more than doubled in the last 200 years. In the poor countries, an incredible miracle has taken place: since W.W.II, life expectancy has almost doubled. This is true for every age level and in almost every country.

Miele: There are published reports that life expectancy has actually declined in the countries of the former Soviet bloc.

Simon: You're absolutely right! There is only one part of the world where life expectancy has decreased—the areas under communism.

Miele: Did that decrease take place in the Soviet period or during the transition to a more market-based economy?

Simon: Murray Feshbach has documented how life expectancy has been going down and mortality going up in those countries, especially among males in the Soviet Union, since the 1950s. That trend has continued through the transition period. This is due to communism, the type of social system they had, and the poverty it produced.

Miele: Economist Amartya Sen, on the other hand, likes to bring up the example of Kerala state in India, which has a more socialist-style economy but has experienced improving health and longevity statistics, along with a more equitable distribution of wealth.

Simon: Yes, that's true. We should also give recognition to the "bare-foot doctors" in Communist China for doing a great deal in raising life expectancy in rural areas. I'm not saying that communism always fails in all ways. But I am saying that the explanation of the demographic and environmental disasters in Eastern Europe was communism and the poverty that it engendered.

I try to see what is good and correct, even in people I don't agree with. There is a marked tendency, however, among those who oppose economic analysis, and even among superb scientists like E. O. Wilson and biologist Robert May—the chief scientific adviser to the British government—to demonize anyone who disagrees with them about any one matter. May, for example, accuses me of dishonesty and having some other agenda simply because I called into question the data that supposedly show an increasing rate of species extinctions.

Miele: Along those lines, you object to the term "cornucopian" being applied to your position. But you do refer to Ehrlich and Hardin as "doomsayers."

Simon: "Doomsayer" is a correct term, "cornucopian" is not. The term "cornucopia" suggests that good things will just come to you like manna falling from heaven. They don't.

Miele: What term would prefer being applied to your argument? Pro-growth?

Simon: I'm certainly pro-growth, pro-human life, pro-liberty, pro-truth. Truth is important to me. Someone asked me the other day, "Should we be down on the doomsayers?" My answer was "Absolutely not!" I have no complaints about anybody who warns us of an impending danger, even if they're wrong, as long as they are truthful. I complain only about those people who knowingly manipulate, exaggerate, or lie for some purpose.

Maybe the term "doomsayer" goes too far in this respect, maybe not. If you say I'm pushing too hard with that term, well maybe you're right. But let's be clear that I object to the term "cornucopian" because it is inaccurate.

Miele: Your specialty is economics, sometimes called "the dismal science." I heard Milton Friedman, whose work I think you respect a great deal, say "there's only one law in economics—there's no such thing as a free lunch." Other economists, especially conservative or libertarian ones, tell us that a key element in economic reasoning is anticipating the unintended consequences of intended actions. At first reading, your arguments in favor of limitless growth can sound like "there's something for everyone at no cost to anyone." How do these varying lines of economic argument jibe? Isn't everything in economics a trade-off?

Simon: First, since the death of Friedrich Hayek, I respect Milton Friedman (and maybe Ted Schultz) above all economists now alive. I also respect him enormously as a human being. I once sent him a copy of one of my books with the inscription, "To Milton Friedman, who is the best man I know or know of."

That said, he's not right on everything. He once was wrong on something on which I was right—the method now used by airlines to offer a cash reward to individuals who agree to accept a seat on a later flight for one on an overbooked flight about to take off. And Milton was man enough to say, "You were right and you're entitled to your 'I told you so.'"

If you call Milton Friedman and say, "Simon says that there may not be any such thing as a free lunch, but the lunch is getting cheaper every year. Do you agree?" he'll say "yes." Friedman is, of course, right that nothing is free. But all that I'm saying is that things get cheaper, and cheaper, and cheaper. Yes, everything is a trade-off; but that doesn't mean that it's all a zero-sum game. The fact that we live longer, eat better, and have more and more of everything every year for less and less effort means that the lunch is getting cheaper and cheaper.

Miele: Does everything keep getting cheaper and cheaper? Daniel Boone probably had an easier time finding peace and quiet than I do.

Simon: Very good. Just the other day I was speaking to a forestry school about the price of timber, which has not fallen the way prices of commodities generally have fallen. Why? Because we're in a transition. Today, firewood is free in Guatemala. You can just go out and cut it and carry it back home. In 50 years, you won't be able to do that. There's a transition period from a subsistence economy to a modern economy. Six hundred years ago in Baku in the Caucasus, you could go to places where oil oozes out of the ground and get oil for free. Now, oil is not free. So let's say that the price of oil was zero back then and it is positive now. That just means it's part of some historical process in which there are certain transitions; but the long-run trend is toward a cheaper-and-cheaper lunch.

If you want tranquillity, you can buy it. You can jet to Alaska or to Timbuktu, or pretty soon, travel into space and get all the tranquillity you want. Most people don't want tranquillity. How many young people do you think want to spend their life in some little town of 200 people?

Miele: Let's turn to your famous bet with Paul Ehrlich. You won, hands down. Now Ehrlich and others argue that you lucked out because the business cycle during the decade chosen for your wager (1980-1990) started with a recession (caused by the world oil shortage) and ended with a recession (possibly brought on by the drop in U.S. real estate prices, resulting from the overage created by Reagan tax cuts, which encouraged the construction of a lot of never-occupied office buildings) What part did the business cycle play in deciding the wager? How general is your confidence?

Simon: If I were Ehrlich and believed what he believed, I would have said in 1990, "Win some, lose some. Let's go again, double or nothing!" If I had lost the bet, and it's entirely possible I could have lost, that's exactly what I would have said. But Ehrlich and the others don't say that because by now they have learned that if you look back historically over the 200 years for which we have data, they would have lost 19 times out of 20. I have not only offered to repeat these wagers, I have offered to expand the wager. I say to these prominent doomsayers, "Pick any measure of material human welfare—life expectancy, infant mortality, automobiles per person, telephones per person, newspapers per person, most especially schooling per person. Pick any country in the world. Pick any year in the future, just as long as it's far enough ahead so that something can happen, not next year, but, say 10 years from now. I'm prepared to bet it will show improvement relative to now."

Miele: Are you familiar with Ehrlich's recent book, *Betrayal of Science and Reason*?

Simon: I am.

Miele: On page 101 he says that he and Steven Schneider have offered you a bet concerning 15 predictions (see list at end of interview). They say, "Simon refused to accept the wager, going back on his original challenge by saying that he will gamble only on "direct" measures of human welfare such as life expectancy, leisure time, and purchasing power" (p.103).

Aren't the things that Ehrlich and Schneider list—temperature, atmospheric levels of carbon dioxide, nitrous oxide and ozone, virgin land, fertile crop land, sperm counts, the gap between the wealthiest and the poorest groups, very important to us?

Simon: I characterize their offer as "switch and bait" because I made it very explicit that my offer deals with measures of material welfare. I'm an economist, not an atmospheric scientist, or a soil scientist, or whatever. I have no predictions regarding the number of ants that will be in my backyard next year, or whether the black squirrels or the gray squirrels will dominate. That's not my trade. Nor do I know that such things matter economically.

What Ehrlich and Schneider have done is to switch from material human welfare to measures of other kinds, which have nothing to do with the subject at hand. They then bait me because I choose not to wager on things that have nothing to do with anything.

Miele: Are you then saying that those things could go exactly as Ehrlich and Schneider have predicted.

Simon: They may, they may not. What I've said to them is—"Look, you're telling me that global warming is bad because it's bad for agriculture. Fine—let's bet on agriculture, or the six or seven other items on their list that can be tied to human welfare." Their response is, "Take it or leave it!"

Let me characterize their offer as follows. I will predict, and this is for real, that the average performances in the next Olympics will be better than those in the last Olympics. On average, the performances have gotten better, Olympics to Olympics, for a variety of reasons. What Ehrlich and the others say is that they don't want to bet on athletic performances, they want to bet on the conditions of the track, or the weather, or the officials, or any other such indirect measure. Isn't that a little bit of a preposterous way to deal with Olympic performance?

One more point. Ehrlich and Schneider want to bet on global warming. Even if one were prepared to make statements about what will happen to the average world temperature, it is utterly unclear as to whether any given change is good or bad. The very same person who is baiting me with this—Steven Schneider—was warning us some 20-odd years ago about the horrors of the impending ice age. Then he was telling us that things were getting cooler and that was bad. Now he's warning us that things are getting warmer and that's bad. Well, it's possible that both of them are taking place and so temperature is at an absolute optimum. If so, it's optimal only because we've adjusted to it for the time being.

Miele: Doesn't one's reaction to the dangers of global warming or of global cooling depend on whether one is a polar bear or a parrot?

Simon: Absolutely. We happen to be neither!

Miele: Then doesn't all this come down to a question of what one is concerned with?

Simon: Of course!

Miele: You read the Garrett Hardin interview [see Chapter 15]. What he and your other critics say is that "Simon looks only at one item at a time, measured by the current market price. What you need to look at are systems and then measure the cost in terms of the overall effect."

Simon: Economists don't only deal with prices. We also deal with their outcomes—things like life expectancy, which is a function of a lot of other things like the price of food and the price of health care. We all certainly care about life expectancy. Economists do deal with market prices and we do say that, for economic purposes, they are measures of scarcity. We don't say they are the only relevant measures of scarcity. If you are a biologist, you may want to study the scarcity of mosquitoes, or some other species. But that concept is only relevant to you as a biologist. The concept that is relevant to us as consumers and as economists is the scarcity of something as measured by what you have to give up to get it—the market price.

Now Hardin says in the interview that the market price doesn't measure the true price. Neither economists, nor Hardin, nor anybody in the world knows what the "true price" is. The "true price" is not a meaningful concept. It is a religious or a metaphysical concept, if anything. The revolution that took place in economics in the 1870s was in understanding the concept of the marginal price. We are not saying that the "true price" of Mother Teresa's time is measured by what she gets paid and the NFL's top quarterback's "true price" is measured by what he gets paid. We simply say that they are the market prices and the market prices measure scarcity.

Miele: That's true by definition, rather than by experiment, isn't it?

Simon: The market price does not represent the intrinsic value. What Hardin is saying is that economists aren't measuring the intrinsic value and that's absolutely right. But that means that he knows what the intrinsic value is. But he doesn't tell you that. He assumes that somehow what he calls the "true price" is some real, underlying value. That is a notion which should run smack against the basis of skeptical thinking. Hardin's "true price" is not an observable phenomenon. It is a metaphysical phenomenon. Or it's one man's judgment that he somehow pretends ought to be the judgment of all humanity.

Miele: In his recent book, Ehrlich draws a connection between what he regards as the antipathy to ecological caution and biblical creationism and other pieces of biological ignorance, as has E. O. Wilson. Christian fundamentalists and certain political conservatives are among the most ardent embracers of your arguments and the most ardent opponents of Hardin and Ehrlich. Is there a religious component in your own views, or those of the ecologists for that matter?

Simon: What do you mean a religious component in my views? Which views?

Miele: The ones you just stated.

Simon: Did I state any views just now? I really didn't. When I say to you that every economist knows that people have been eating better, that is not a view. That's a fact. It's a fact that people have been eating better and it's a fact that almost every agricultural economist knows it. I have mostly limited my answers to statements of fact and explanations of them.

Any discussion of religious views is inappropriate here. Anything that I say must stand by itself on the basis of its scientific content. Herman Daly [an economist cited by Garrett Hardin as understanding the "ecological point of view"], on the other hand, in a book of his will say that we should be saving energy because God wants us to.

Miele: E. O. Wilson characterizes the argument of creationist anti-ecologists as the belief that, since humanity is somehow exempt from the laws of evolution and genetics that govern all other species, we therefore don't have to worry about anything because God will provide for us. You sometimes talk about human ingenuity as if it can play such a God-like role. Or do you draw a distinction between the two?

Simon: Nothing in that previous statement resembles anything that I would say. I would say that human ingenuity is a fact, you can measure it objectively in terms of patent records and the number of inventions.

Miele: But you would say that *Homo sapiens* is subject to the same rules of genetics and evolution that govern all other species on Earth?

Simon: I certainly would not say that; I don't know anything about it. Frank, this is very relevant to the principles of SKEPTIC magazine. It is very important that everything that I say be reasonably observable, testable, and within my area of competence. I do my damnedest never to speak publicly about anything that might allow someone to accuse me of trying to speak like an expert on something when I'm not.

After he got the Nobel Prize, Hayek wrote a piece about why economists should not be given Nobel prizes. You have to get a Nobel prize to be able to say that. Why? Because when you get a Nobel prize the media puts a microphone in front of you and asks you about stuff you know nothing about, and you answer. It is my business never to provide such answers. And that is why I have been able to survive all these years of controversy without anyone finding me in gross error in the things that I have said or written. There is no statement, with about two exceptions, that I have written in the past 30 some-odd years that I could not write again today and feel comfortable with.

When you start asking me about genetics and God and such things, I'm an expert on neither. Luckily, this comes congenially to me because I have always hated opinions.

Miele: Well that's a good skeptical position. You do, however, in your book cite both published works and a personal communication from cosmologist Frank Tipler, whose worked is discussed in SKEPTIC Vol. 3, #4.

Simon: Yes, but I cite Tipler because he sent me a minute specimen of some rare metal that transmutes itself into copper as living proof that Ehrlich, et al. were wrong when they dismissed my argument that the amount of copper on Earth is not fixed and that you can make copper from any other metals. That's where I cite Frank Tipler, not his views on God, immortality, or religion.

Miele: You do cite Tipler's views on infinity, that "because we can increase the stock of information without limit, there is no need to consider our existence finite" (*Ultimate Resource 2*, p. 65).

Simon: That's correct, but that has nothing to with God or whatever.

Miele: In Tipler's mind it does!

Simon: "In Tipler's mind it does!" How did I get to be Tipler? Come on!

Miele: Then let me turn to something that is in your bailiwick. I think I first encountered the argument in an editorial by fellow economist Thomas Sowell that everyone on Earth could live in a three-bedroom house in Texas, or we could fill the Earth to the population density of the Netherlands.

Simon: I recalculated the numbers and he had it exactly right.

Miele: What Hardin and Ehrlich would reply is that those people could physically fit into Texas or be packed at the density of the Netherlands, but the real question requires you to calculate the total draw that contemporary Texans or the Dutch make on the resources of the Earth, including the labor of all the rest of us. Do you think everyone on Earth could live at the affluent level of consumption of the typical Dutch or Texas middle-class family?

Simon: Absolutely. With each year, humanity gets closer and closer to achieving that. One hundred years from now, an even greater percentage of humanity will have reached that level of prosperity. Humanity is moving exactly in that direction and will move ever further in that direction, without limit!

Miele: In your book on population you discuss your earlier book on overcoming depression and describe how your change in mood was related to your own change of, if not opinion, then observation of the question of population and prosperity.

Simon: My economic work was utterly independent of my mood. Rather, the change in mood was somewhat affected by my economic findings.

Miele: The reason I ask is because when I interviewed Hardin I told him that I couldn't imagine him or Ehrlich writing books such as you have on *Overcoming Depression* or *Start Your Own Mail Order Business*. There seems to be an element of depression or pessimism inherent in the arguments of the ecologists.

Simon: Well maybe. That's interesting. Except that as individuals, they are at least as light-hearted as I am. That's an astonishing thing about them. Just as they advise everybody else not to drive cars—while they themselves are jetting and driving all around the world—by the same token, they are uttering these doomful propositions about the world, but it's not really reflected in their own behavior. However, Ehrlich has said that he has had only one child because of his commitment to population control. I'm sorry he didn't have four.

Here is where Ehrlich and I differ. I think that if he had four kids, the world would be a richer place. The world is certainly a richer for him being here. He tells *The Wall Street Journal* that the world would be better off if I wasn't around. I want more Ehrlichs, more Hardins, and their children.

Miele: You admonish your reader on page 10, "If no data could falsify your belief, then your position is a matter of metaphysical belief, or theology, rather than science." On page 554 you state, "the evidence persuades me that...eugenics policies have little or no rationale—and they would be unacceptable to me even if there were evidence in their favor." Isn't that an expression of a religious or an ethical, rather than a scientific belief by your own definition.

Simon: Yes, I express ethical views occasionally. I hope that when I do, I label them as such.

Miele: Do you believe that all people are a net plus to society and the world?

Simon: No. I'm only saying that before a child is born, on average, the odds are that child will contribute more than that child takes, no matter whose child it is.

Miele: What do crack-addicted babies, anencephalic babies, or terminally-ill patients, who consume vast resources to be kept artificially alive, contribute?

Simon: Our measures of value depend upon people being consumers of value. The basic assumption that most of us make is that human life itself is a good thing. Even Ehrlich would agree; Garrett Hardin, maybe not when he says that we should only have half as many or a quarter as many people as we now have. But for most people, one of our underlying assumptions, which I'm prepared to make explicit, is that we are concerned with material human welfare. Without human beings to experience it, there cannot be human welfare.

We will always have human problems—we're talking on average. We always make blunders, we always have disasters, and there are always short-term ups and downs. I'm interested in the long term where the data are clear. All I say is that for every 99 steps backward, we take 100 steps forward.

Miele: Is there anything you would like to add or to correct?

Simon: Only to emphasize what we started with. I want to raise the question of the editorial policies of SKEPTIC magazine. It's an admirable venture. I hope that it will be carried forth with devotion to its central tenets, rather than engaging in controversy for controversy's sake.

Miele: Thank you for taking some of your free time, which I'm sure you'll agree with me, is a finite and scarce resource.

Simon: It's the only thing we've got that's finite!

EHRLICH AND SCHNEIDER'S LIST OF 15 ITEMS "PERTAINING TO MATERIAL HUMAN WELFARE" THEY PREDICTED WOULD WORSEN OVER THE NEXT DECADE AND ON WHICH SIMON DECLINED TO BET ARE:

1. The three years 2002-2004 will on average be warmer than 1992-1994.
2. There will be more carbon dioxide in the atmosphere in 2004 than in 1994.
3. There will be more nitrous oxide in the atmosphere in 2004 than 1994.
4. The concentration of ozone in the lower atmosphere (the troposphere) will be greater in 2004 than in 1994.
5. Emissions of the air pollutant sulfur dioxide in Asia will be significantly greater in 2004 than in 1994.
6. There will be less fertile cropland per person in 2004 than in 1994.
7. There will be less agricultural soil per person in 2004 than 1994.
8. There will be on average less rice and wheat grown per person in 2002-2004 than in 1992-1994.
9. In developing nations there will be less firewood available per person in 2004 than in 1994.
10. The remaining area of virgin tropical moist forests will be significantly smaller in 2004 than in 1994.
11. The oceanic fisheries harvest per person will continue its downward trend and thus in 2004 will be smaller than in 1994.
12. There will be fewer plant and animal species still extant in 2004 than in 1994.
13. More people will die of AIDS in 2004 than in 1994.
14. Between 1994 and 2004, sperm counts of human males will continue to decline and reproductive disorders will continue to increase.
15. The gap in wealth between the richest 10% of humanity and the poorest 10% will be greater in 2004 than in 1994.

References

Daly, H. 1977. *Steady-State Economics*. San Francisco, CA: Freeman.

Ehrlich, P. and Ehrlich, A. 1996. *The Betrayal of Science and Reason*. Washington, D.C.: Island Press.

Feshbach, M. and Friendly, A. 1992. *Ecocide in the Soviet Union*. New York, NY: Basic Books.

Feshbach, M. 1995. "Mortality and Health in the Soviet Union." In Simon, J. 1995. *The State of Humanity*. Boston: Basil Blackwell.

Hardin, G. 1993. *Living Within Limits—Ecology, Economics, and Population Taboos*. New York, NY: Oxford University Press.

Johnson, D. G. 1970. "Famine." in *Encyclopedia Britannica*.

Leikind, B. 1995. "Can Science Prove God? Frank Tipler's The Physics of Immortality." *Skeptic*, Vol. 3, No. 4.

McLeod, R. 1997. "Religious Leaders Add Activist Voice to Population Debate." *San Francisco Chronicle*. February 28.

Miele, F. 1996. "Living Within Limits and Limits on Living. Garrett Hardin on Ecology, Economy, and Ethics." *Skeptic*, Vol. 4, No. 2.

National Research Council, Committee on Population, and Working Group on Population Growth and Economic Development. 1986. *Population Growth and Economic Development: Policy Questions*. Washington, D.C.: National Academy Press.

Schultz, T. 1984. *Investing in People*. Chicago, IL: University of Chicago Press.

Sen, A. 1987. *The Standard of Living*. Cambridge, MA: Cambridge University Press.

Sen, A. 1992. *Inequality Reexamined*. Cambridge, MA: Harvard University Press.

Shermer, M. 1995. "Hope Springs Eternal: Dr. Tipler Meets Dr. Pangloss." *Skeptic*, Vol. 4, No. 2.

Simon, J. 1990. *Population Matters: People, Resources, Environment, Immigration*. New Brunswick, NJ: Transaction Press.

Simon, J. 1993. *Good Mood: The New Psychology of Overcoming Depression*. Chicago, IL: Open Court.

Simon, J. 1995a. "Earth's Doomsayers Are Wrong." *San Francisco Chronicle*, 18 May.

Simon, J. 1995b. *The State of Humanity*. Boston, MA: Basil Blackwell.

Simon, J. 1996. *The Ultimate Resource 2*. Princeton, NJ: Princeton University Press.

Thorne, K. and Tipler, F. 1995. "A Cosmological Dialogue on The Physics of Immortality." *Skeptic*, Vol. 4, No. 2.

Wilson, E. O. 1996. *In Search of Nature*. Washington, D.C.: Island Press.

Frank Miele

CHAPTER 17

HOW CLOSE ARE WE
TO CLONING TIME?

AN INTRODUCTION TO THE SCIENCE
AND ETHICS OF HUMAN CLONING

"The Commission concludes that at this time it is morally unacceptable for anyone in the public or private sector, whether in a research or clinical setting, to attempt to create a child using somatic nuclear transfer cloning."
—*Report on Cloning Human Beings*, National Bioethics Advisory Commission.

"What nonsense, what utter nonsense, to think we can hold up our hands and say, 'Stop.' Human cloning will take place, and it will take place in my lifetime. I don't fear it. I welcome it."
—Senator Tom Harkin, *Senate Committee Hearings on Human Cloning.*

The very phrase "genetic technology" linking genetics, "the science of heredity," with technology, "the practical application of knowledge, especially to engineering purposes," is the stuff of which horror stories and science fiction are made. For most of human history, life has been seen as a gift, perhaps the greatest gift, of God or the gods. But since Mendel first started breeding peas and especially since Watson and Crick determined the molecular structure of DNA, the mysteries of life have become increasingly demystified. The emerging field of genetic technology has given us new forms of life, the ability not only to detect but to cure genetic disorders, and even the ability to build even more sophisticated biological weapons, including the spectre of ethnically targeted bioweapons. The cloning of Dolly the sheep suggests we are close to cloning a human. To many that means "taking a major step into making man himself simply another one of the man made things" so that "human nature becomes merely the last part of nature to succumb to the technological project which turns all of nature into raw material at human disposal" (Kass, quoted in *National Bioethics Advisory Board* [*NBAC*], 82).

Who should control human reproduction—God, society, tradition, the state, the individual? Is cloning a human next? If so when? What are the reasons for human cloning and what are the risks? How close are we to cloning time? This overview of the science behind cloning and ethical questions it raises. It is followed by my interview of Dr. Richard Seed, "the man who would be cloned," who has become the point man for human cloning ever since he announced he plans to clone himself.

FROM CUTTINGS TO CLONING: SOME SCIENTIFIC BACKGROUND

Getting at the deeper issues involved in cloning and genetic engineering requires brief overview of the ever growing field of genetics. The structure of the DNA molecule as a twisted ladder with AT and GC base pair bonds as the rungs has provided a completely materialistic basis for the defining properties of living things:

> •a self-directing information code that has a structure which
> is maintained in stable form,
> •and so is able to serve as a model to replicate itself,
> •yet is susceptible to change that is transmissible.

Sexual reproduction, differentiation, and development into the various cells of a complex organism, growth from single cell to fetus to newborn to adult, disease and repair, even senescence and death, are each day proving more understandable in terms of the underlying biochemistry. An important point is that the duplication of DNA is not always exact. The sequence of base pairs occasionally contains a "typo," which provides for the introduction, accumulation, and transmission of mutations. The on-going frequency of such typos is actually rather high, about 1 per 100,000 base pairs. We now know that the DNA code acts like a program which contains its own "spell-checker" that detects most of these errors and corrects them (Weaver and Hedrick, 183). In fact, the spell-checker itself has evolved so as to regulate the number of errors it lets through. Feed that controlled variation into the Darwinian selection machine and we can understand not only the biochemical basis of the life of an individual, but the evolution of more complex life forms over time.

Viewed from this perspective, the bodies, sensory organs, and behaviors of complex organisms appear as vehicles for transmitting what Richard Dawkins so aptly dubbed "selfish genes." The distinction between non-living things and living things becomes a matter of increasing information—organized complexity, not categorical difference. Crystals mark the end point on the non-living side, viruses the living side, with the area in between becoming increasingly gray and muddled.

In discussing cloning it is especially important to remember that:

1. The complete genetic code for each and every cell in our body—the genome—is contained in the DNA that composes the chromosomes of every cell of our bodies.
2. There are 23 chromosomes in each of the human germ line cells (sperm or eggs), and 23 pairs of chromosomes in each of the somatic cells.
3. We now have the technology to recognize, remove, or replace sequences of DNA base pairs, to use them as a trigger to stimulate biological processes that are linked to them in some manner, to duplicate them, or to use them to restart the process of development and gene expression.

ALL CLONES ARE NOT THE SAME

The word "clone" derives from the Greek for a twig, and horticulturists have been taking cuttings and growing new plants from them for centuries. The word came into current usage when the renowned British biologist J. B. S. Haldane suggested in 1963 that it would soon be possible to create genetic duplicates of plants, animals, and even humans. Nobel laureates Joshua Lederberg and James Watson also discussed the prospects of future cloning (Kolata, 70-73; Pence, 1998, 1-11). Numerous science fiction books and movies then took up the term and popularized it. The two best known to the general public are David Rorvik's *In His Image,* and the legal brouhaha that followed its publication of as a work of non-fiction, which resulted in the publisher (but not the author) stipulating to its being "a hoax and a fraud," and Ira Lewin's novel *The Boys From Brazil,* about Dr. Josef Mengele's program to clone Adolf Hitler, later made into a movie starring Gregory Peck and Sir Laurence Olivier.

According to the National Bioethics Advisory Commission (NBAC), a clone is "a precise genetic copy of a molecule, cell, plant, animal, or human being" and cloning refers to any of a number of "established technologies that have been part of agricultural practice for a very long time and currently form an important part of the foundations of modern biological research" (25). Regenerating complete plants from cuttings is commonplace in agriculture. Simple animals such as flat worms can regenerate completely from a small segment, and even vertebrates such as lizards regenerate tails and limbs. While this type of regeneration does not take place in mammals, millions of human clones—identical (monozygotic) twins—walk the earth today. Each pair of twins developed from a single ovum that was fertilized by a single sperm and then at some point during the pregnancy split into two (or more). If the split is not total, it produces conjoined (what used to be termed "Siamese") twins. Identical twins are usually very similar in their bodies and their behavior, but they are not "identical persons." Conjoined twins certainly aren't, and Dolly-type clones wouldn't be either.

The biotechnology revolution has brought four new types of cloning: (1) Molecular Cloning; (2) Cellular Cloning; (3) Blastomere Cloning, the technique used to clone the monkeys, Netty and Ditto, born in 1997; and (4) Nuclear Transplantation Cloning—i.e., Hello Dolly!

Molecular cloning uses a host bacterium to duplicate a segment of DNA that has been found to produce a biologically important substance (e.g., insulin for treating diabetes). Cellular cloning makes copies of particular somatic (not germ line) cells. It is extremely useful in testing the effects of new drugs on the cells of a particular part of the body. (An example would be cloning kidney cells to test the effects of EPO, which is used to treat the anemia associated with kidney dialysis.) Neither of these methods, which are commonplace in the biotechnology business and a vital part of modern medicine, produces anything like the sci-fi clone that grows from a fertilized egg into a complete genetic but time-lagged duplicate of a sheep, goat, or another Hitler.

Figure 17.1 shows how we make babies the "good old fashioned way." The chromosomes in the sperm of the father join with those in the egg of the mother to produce an embryo that develops into a baby sheep, goat, or human. If however it splits during development, identical twins are produced.

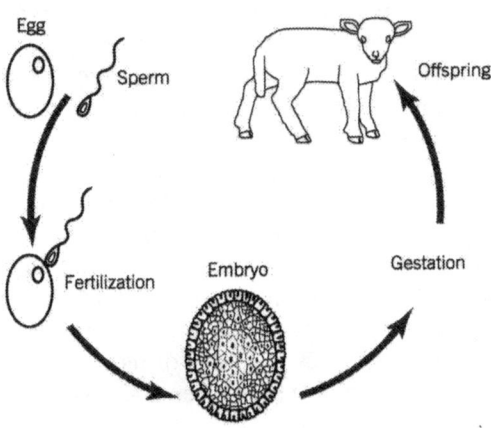

Figure 17.1: Old Fashioned Sexual Reproduction.

Biotechnology now allows us to perform the fertilization in vitro, induce splitting, and then implant the offspring into surrogate mothers. That technique, termed blastomere separation, has become an effective method of producing two or more genetically identical sheep or goats (which is particularly valuable when used on genetically engineered animals that produce important biochemicals). As shown in Figure 17.2, however, the separation must be induced early in the development sequence. Once what is termed the blastocyst stage (in which the inner cell mass has developed) is reached, it is no longer possible to produce clones.

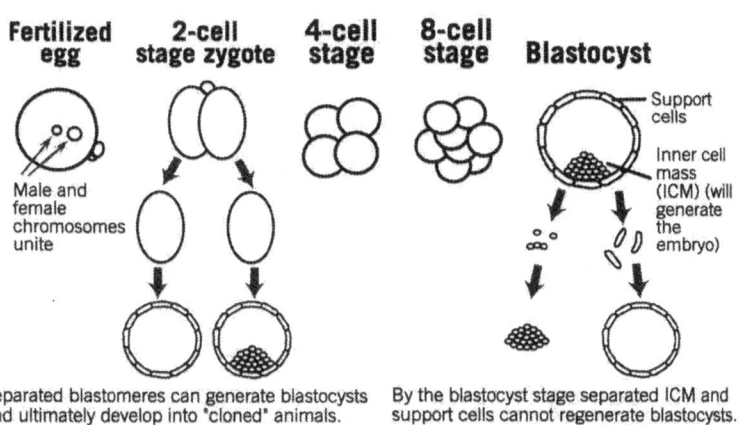

Figure 17.2: How far in the developmental process does our DNA retain totipotency—the potential to generate a complete new organism?

While blastomere separation is a long way from the clones of sci-fi, it illustrates the concept of totipotency, which is critical not only in producing clones of adults, but to Richard Seed's Holy Grail of human rejuvenation. The blastomeres in Figure 17.2 are totipotent, that is, they have the potential to generate a complete new organism; the blastocysts in Figure 2 are not. The key scientific issue that underlies the entire technology of cloning is the question of how long the DNA remains totipotent. From the moment of conception, the fertilized egg begins to divide and develop into different types of cells and tissues, such as the muscles in our arms and legs, the liver, kidneys, and especially the neurons in the brain.

How far in the developmental process does our DNA retain totipotency? DNA in the germ line cells (sperm and eggs) remains totipotent throughout life. But what about the somatic cells? All the somatic cells contain the same DNA, but the particular genes that are activated determine which type of cell develops. A potential neuron, for example, must turn on the neural genes, but turn off the genes that activate the development of other somatic cells. Likewise for a liver cell or a muscle cell.

How long does the DNA in a cell in the liver or the kidney or the brain retain the ability to repeat the entire developmental sequence and produce not just another liver, kidney, or brain, but a complete genetically identical delayed twin? Is the duration of totipotency the same for all somatic cells? Does it vary among species, decreasing with biological or neural complexity? Does it vary among individuals within a given species? Is it itself genetically determined? Is it subject to selective pressure? The answers to those questions will determine whether human cloning remains science fiction or becomes science fact.

Experiments with tadpoles as far back as the 1950s and 1960s demonstrated that, at least in some amphibians, somatic cells retained totipotency. As shown in Figure 17.3, if the nucleus of a somatic cell could be injected into a recipient ovum in which the nucleus had been deactivated, a small number would develop into adult frogs.

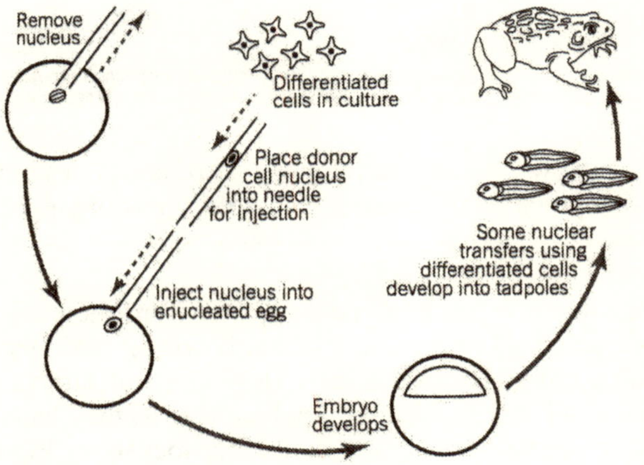

Figure 17.3: In some amphibians, somatic cells retain totipotency. When the nucleus of a somatic cell is injected into a recipient enucleated ovum, a small number develop into adult frogs.

Other experiments successfully transplanted the nuclei of adult frogs. A very small number (4%) developed into tadpoles, but none of these became adult frogs, which suggested that potency might be declining as a function of donor age (or degree of cellular differentiation). The state of knowledge circa 1978 suggested that cloning a man from an adult somatic cell was a long shot at best. So Rorvik's *In His Image* had to be science fiction, not science fact.

Research in the 1990s, however, demonstrated that most somatic cells do retain some degree of totipotency. Then with the technique of using an electric pulse to fuse nuclei of somatic cells with unfertilized eggs, the door for cloning a human suddenly and loudly swung open. Figure 17.4 shows the method by which Ian Wilmut produced Dolly. Donor cells from the mammary gland cell cultures of an adult sheep were placed next to an egg cell from which the nucleus had been removed. A shot of electric current fused the donor and recipient cells. Most importantly, it set the DNA differentiation program back to time zero.

Of 277 successful fusions, however, only 29 (11%) developed to the blastocyst stage. When these were transferred into surrogate mothers, only 1 produced a live, bleating little lamb.

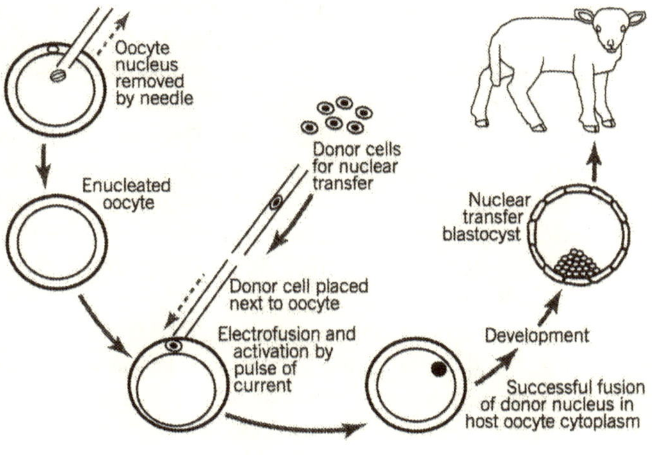

Figure 17.4: The method by which Ian Wilmut set the DNA differentiation program back to time zero and produced Dolly. Donor mammary cells of an adult sheep were placed next to an enucleated egg cell. A shot of electric current fused the donor and recipient cells.

SCIENTIFIC QUESTIONS RAISED BY DOLLY

Dolly marked the first time a fully developed mammal had been cloned by somatic nuclear transfer. She was a one in 277 shot, not good odds at any race track. And a number of important scientific questions remain before human cloning becomes a sure thing:

- Was Dolly really produced from a fully differentiated mammary gland cell or from a less differentiated mammary stem cell? Wilmut's experiment might not really have proven that it is possible to turn back the program to time zero for fully differentiated cells.
- Just how old was Dolly at birth? Somatic cells age each time they divide during the normal course of life. This is probably because the ends of the chromosomes, termed telomeres, are truncated on each cell division. For germ line cells, the enzyme telomerase helps the telomeres restore their full length. But this is not the case for somatic cells. While there is telomerase in the recipient nucleus, will it protect and restore the ends of the donor chromosomes? Were Dolly's cells really those of a newborn little lamb or were they already those of an old ewe when she emerged from the womb? Recent evidence says that Dolly (or at least her genetic material) was 6 years old at birth, because her chromosomes are shortened.
- Normally, mutations in somatic cells affect only that cell and the cells it produces when it divides. Any mutations that occurred in the mammary cells were transmitted into Dolly. Are somatic cells more subject to mutation or to more lethal mutations than germ line cells? Was Dolly born with a heavy genetic load?
- Embryonic gene activation occurs at the 8-16 cell stage in sheep, but at the 4-8 cell stage in humans. Is the window of opportunity in humans too narrow for successful somatic nuclear transfer? Would the success rate in humans be substantially below Dolly's 1 in 277?
- By definition, cloning reduces genetic variation. Reduced genetic diversity has already produced problems in inbred strains of wheat and corn, leaving them dangerously susceptible to killer strains of viruses and bacteria. Purebred dogs, while remarkable not only in their physical attributes, but also their behavior, are subject to a host of genetic

disorders. (The solution is to breed back to other strains, which agriculturists, but not dog fanciers, realize and practice). Clones would all share the same genetic risk factors. A nation of clones could be wiped out by a single virus.

- While the thought of cloning another Michael Jordan or Albert Einstein may be appealing, a little reflection shows that neither of them (or anyone else of prominence) was produced by cloning. The whole evolutionary advantage of sexual reproduction is that it increases genetic variability by continually throwing up new combinations of genes to face new environmental challenges. Whether investing in the stock market or descendants, diversification is a wise long term strategy in an uncertain world.

RELIGIOUS, ETHICAL, AND MEDICAL CONCERNS

Before reaching its conclusion urging a moratorium on human cloning, the NBAC examined not only the scientific, but also the ethical, legal, and even religious implications of cloning. Virtually everyone agrees that the degree of medical risk involved needs to be considered when human cloning is contemplated. But they disagree on the extent to which they believe that risk will remain above that associated with other, acceptable methods of reproductive technology (such as in vitro fertilization), or even ordinary reproduction.

The religious arguments against human cloning are as varied as the religious doctrines from which they derive. Their only consistency is that those religions or religious persuasions (e.g., Roman Catholicism) that have adamantly opposed other means of giving individuals control of reproduction adamantly oppose cloning, while those that have been more open to such methods (e.g., Reform Judaism) are more open to cloning. The NBAC could only conclude that "the wide variety of religious traditions and beliefs epitomizes the pluralism of American culture—there is no single 'religious' view on cloning humans, any more than for most issues in biomedicine" (72). The Commission also weighed ethical arguments, as opposed to purely religious arguments against cloning, including the possibility of a diminished sense of individuality and personal autonomy, degradation of the quality of parenting and family life, whether parents would be tempted to seek excessive control over cloned children, and the re-opening of the door to eugenics. They compared these arguments with those in favor of cloning, including protecting personal choice, maintaining privacy, freedom of inquiry, and encouraging the development of powerful new technologies. They concluded that their efforts were hamstrung by the fact that "neither moral philosophers nor religious thinkers can agree on the 'best' moral theory; indeed, they often cannot even agree on the practical implications of any single theory" (76-77).

The most ardent academic opponent of human cloning is bioethicist Leon Kass, of the University of Chicago. Kass testified before the NBAC and elaborated the anti-human cloning position against political scientist James Q. Wilson's measured defense in the "opposing viewpoints" book, *The Ethics of Human Cloning* (Kass and Wilson, 1998), published by the American Enterprise Institute think tank. Kass, himself a former molecular biologist, builds his argument against human cloning on the "wisdom of repugnance": "the emotional expression of deep wisdom beyond reason's power fully to articulate it." Kass concedes that "some of yesterday's repugnances are today calmly accepted" although he qualifies the concession by noting: "though, one must add, not always for the better" (18). It should come as no surprise, then, when Kass places the specific issue of cloning within the broader conflict between societal and individual control of reproduction:

> Cloning turns out to be the perfect embodiment of the ruling opinions of our new age. Thanks to the sexual revolution, we are able to deny in practice, and increasingly in thought, the inherent procreative teleology of sexuality itself. But, if sex has no intrinsic connection to generating babies, babies need have no necessary connection to sex. Thanks to feminism and the gay rights movement, we are increasingly encouraged to treat the natural heterosexual difference and its preeminence as a matter of "cultural construction." But if male and female are not normatively complementary and generatively significant, babies need not come from male and female complementarity. Thanks to the prominence and the acceptability of divorce and out-of-wedlock births, stable, monogamous marriage as the ideal home for procreation is no longer the agreed-upon cultural norm. For that new dispensation, the clone is the ideal emblem: the ultimate "single-parent child."
>
> Thanks to our belief that all children should be wanted children (the more high-minded principle we use to justify contraception and abortion), sooner or later only those children who fulfill our wants will be fully acceptable. Through cloning, we can work our wants and wills on the very identity of our children, exercising control as never before. Thanks to modern notions of individualism and the rate of cultural change, we see ourselves not as linked to

ancestors and defined by traditions, but as projects for our own self-creation, not only as self-made men but as man-made selves; and self-cloning is simply an extension of such rootless and narcissistic self-re-creation. (emphasis in the original, 8-9).

Compared to Kass, James Q. Wilson provides only the most qualified defense of controlled human cloning in the AEI volume. Much stronger cases are made by such cloning advocates as the philosopher Gregory A. Pence in his book *Who's Afraid of Human Cloning?* and Lee Silver, a professor of Molecular Biology, Ecology, Evolutionary Biology, and Neurosciences at Princeton University, in his book *Remaking Eden: Cloning and Beyond in a Brave New World*. Pence had already begun his book before the NBAC was convened. Attending the meetings, Pence was disappointed that none of the commission members "was willing to defend human cloning in any way" and that "professionalism in bioethics meant that each side had to be defended with logic and passion" (xi-xii). Pence shows that the very same arguments, including medical risk, have all been made against other reproductive technologies, most recently, in vitro fertilization. Technical advances and market acceptance have disposed of all of them.

Pence also edited the opposing viewpoints reader, *Flesh of My Flesh*. It includes essays against by Kass; by theologian George Meilander that cloning cheapens God's gift of human reproduction (39); and by bioethicist George Annas that cloning is replication, not reproduction, and would therefore cross a natural boundary of kind rather than degree. Annas opined against the NBAC and other government commissions that "popular notions of cloning derive from science fiction books and films that have more to do with cultural fantasies than actual scientific fact" and that they were "wrong to disregard the lessons from our literary heritage on this topic, thereby attempting to sever science from its cultural context" (80). The book also includes an essay by philosopher Philip Kitcher, who would only allow cloning when there is no other way to produce genetically related offspring, as for example for a lesbian couple; as well as essays by Stephen Jay Gould and by Richard Lewontin, debunking fears of genetic determinism.

In *Remaking Eden* Silver argues that rather than a technological curse, genetic engineering will provide all the cures for cloning, as well as ordinary reproduction. He points out that good old-fashioned reproduction produces plenty of errors and has a lower overall success rate than some modern reproductive technologies. Critics of cloning and other forms of reproductive technology can't have it both ways. If medical risk is a serious enough moral concern to demand banning a given reproductive technology, then the existing risks of ordinary reproduction must produce an equal and opposite need to develop methods of reducing or eliminating those risks.

Pro-cloners dismiss the concerns about reduced genetic variability by pointing out that cloning will no more replace "making babies the good ole fashioned way" than has any other reproductive technology. As Richard Seed notes in the interview that follows in Chapter 18, the applications to his human cloning clinic have come from infertile couples, not influential billionaires. Further, Seed argues that advanced reproductive technology opens the door for unlimited genetic variability. He dismisses all the ethical and religious objections against human cloning, and considers the medical risk the only valid argument. Increasingly, however, that risk is proving to be real. Cloning experiments on mice, goats, cows, and monkeys, while admittedly still in their infancy, nonetheless already have produced some serious medical problems.

Reuters (1999) summarized a report published in the May 1st issue of the British medical journal, *The Lancet*, on the sudden death of a cloned calf that led French researchers to suspect the cloning process itself causes long-term health problems. From birth, the calf's white blood cell count fell rapidly. There was a quick decline in hemoglobin—the oxygen-carrying pigment found in red blood cells. Iron supplements did not improve the calf's condition, and it died of severe anemia at the age of 51 days. Postmortem examination revealed a seriously underdeveloped thymus, spleen, and lymph nodes—organs that play a key role in the immune system.

The report noted that "cloning has a relatively high rate of late abortion and early postnatal death," with current neonatal death rates among cloned animals approaching 50%. The researchers concluded that cloning may pose an inherent risk to long-term health, noting that their observations "should be taken into account in debates on the effective application of reproductive somatic cloning to human beings."

The Oregon Regional Primate Research Center produced Netty and Ditto through the process of blastomere cloning. But they have not been able to clone a monkey successfully by somatic nuclear transfer. Their attempts have resulted in death of the clones and sometimes even their pregnant mothers. The Oregon team's experiments have produced placental abnormalities, abnormal swelling, three to four times the normal rate of umbilical cord problems, and severe immunological deficiencies. These abnormalities are similar to those found in experiments on disrupting the phenomenon of gene imprinting—the switching on or off of genes based on whether they come from the mother or the father. Since clones have only one parent, faulty imprinting may be the cause (Weiss, 1999).

Recent evidence from the Roslin Institute (*Chronicle News Service*, 1999) shows that Dolly's telomeres are 20 percent shorter than is normal for a 3-year old sheep. Cloned from a 6-year old ewe, Dolly's shortened telomeres suggest she was biologically six at the time of her birth. Each time the DNA unwinds, the telomeres (a small piece at the end of the chromosome that may help to seal and then release the two strands) are truncated. Like a barber or a loan shark, the aging process always "takes a little bit off the top." Eventually, like Humpty Dumpty, the double helix can't be put back together again. But are telomeres the mechanism of aging, or just one of its products (like gray hair), or its markers (like worn-down molars)? To complicate matters further, Dolly has given birth to two lambs. If Dolly was prematurely aged at birth were her offspring as well? Only time and continued observations will tell. But the latest research on sheep is baaaaaad news for Richard Seed's goals of human cloning and rejuvenation. Instead of a fountain of perpetual youth, the result of repeated DNA reprogramming may be to turn out the techno version of the Struldbruggs of Swift's *Gulliver's Travels*—they live forever, but continually age, ending up a bunch of blind, deaf, slobbering invalids, surely new grist for the sci fi clone mill.

A MATTER OF PRUDENCE

The issues of cloning and other reproductive technologies cannot be decided by religious or ethical argument because there simply is no agreed upon religion or ethical system. Experts reach totally opposite conclusions. Kass, an esteemed bioethicist who is knowledgeable in the biosciences, argues vehemently that cloning is but the last step in the breakdown of the protection given by society to the normal family. Kitcher, an equally distinguished and erudite ethicist, is troubled by human cloning but thinks it would be justified for the purpose of building what Kass would find a most un-normal family. Pro-cloners Silver and Seed have faith that greater knowledge and technical expertise will solve all problems.

Human cloning is really a matter of prudence, defined as "proceeding from caution and good judgment." Ian Wilmut and the Roslin Institute originally conducted their sheep cloning experiments as a way to decrease the costs to support the institute's other ongoing biogenetic research. When the media raised the issue of human cloning, they backed away because they saw the subject as only getting in the way of advancing scientific knowledge and future profits. The NBAC moratorium was an attempt to assuage the public fears raised by that media coverage. As human cloning advocate Richard Seed points out, the NBAC prohibition on human cloning has no teeth. Congress has taken no binding action. Even it if did, how could they ban all forms of cloning? The slightest change in procedure could legally qualify as being something other than somatic nuclear transfer. If banned in the U.S. (and elsewhere), it would only go to countries eager to attract capital and technology.

The opening quotations from the NBAC and Senator Harkin are both correct. Given the medical risks and the scientific unknowns, human cloning "just wouldn't be prudent at this time," as former President George Herbert Walker Bush used to say. Seed's Human Cloning Clinic or some other group may announce that they have cloned a human, even before you read this article. But if so, it will only be like landing a man on the moon, a sensation rather than a scientific breakthrough. The latter will come from innumerable experiments, large and small, some ballyhooed, most unreported by the mass media, that explore the underlying issue of genetic development—how does the DNA code direct itself? Before Richard Seed's goal of rejuvenation becomes reality, we will have to answer that question precisely for each and every organ in the body, or at least the master genes that code for them.

REFERENCES

Chronicle News Service. 27 May 1999. "Sheep clone was born old, study says" p. A-1

Kass, L. and J. Q. Wilson. 1998. *The Ethics of Human Cloning.* Washington, DC: AEI Press.

Kolata, G. 1998. *Clone: The Road to Dolly and the Path Ahead.* NY: Norton.

National Bioethics Advisory Commission (NBAC). 1997. *Cloning Human Beings: Report and Recommendations.* Rockville, MD.

Pence, G. E. (ed.). 1998a. *Flesh of My Flesh: The Ethics of Cloning.* New York: Rowman & Littlefield.

Pence, G. E. 1998b. *Who's Afraid of Human Cloning?* New York: Rowman & Littlefield.

Reuters Health News. 30 April 1999. Summary of article in *The Lancet*, Vol. 353: 1489-1491.

Silver, L. M. 1997. *Remaking Eden: Cloning and Beyond in a Brave New World.* NY: Avon.

Weaver, R. F. and P. W. Hedrick. 1997. *Genetics.* Chicago, IL: William C. Brown Publishers. 3rd Edition.

Weiss, R. 1999. "Clone Defects Point to Need for 2 Genetic Parents." *Washington Post.* May 10, 1999, P. A1.

Frank Miele

CHAPTER 18

THE MAN WHO WOULD BE CLONED

AN INTERVIEW WITH DR. RICHARD SEED, DIRECTOR OF THE HUMAN CLONING FOUNDATION

How did physicist Richard Seed become the point man in the race to produce the first human clone? Back in 1978, science writer David Rorvik published *In His Image: The Cloning of a Man*, in which he claimed to be part of a team that had successfully produced an exact genetic replica of a mysterious, reclusive Howard Hughes-like billionaire named "Max." British embryologist J. Derek Bromhill sued Rorvik's publisher, J. B. Lippincott, on the grounds that Rorvik had cited his scientific work so as to imply Bromhill was involved. After a series of court battles, Lippincott paid damages to Bromhill and the court ruled *In His Image* to have been "a fraud and a hoax."

Following the Rorvik affair, scientists and science writers stayed away from the subject of human cloning, though it remained fertile ground for science fiction. The received wisdom was that cloning only worked in simpler animals such as frogs. That all changed in 1996 when Ian Wilmut and a team at the Roslin Institute in Scotland published an article in *Nature*, Britain's most prestigious science journal, describing how they had cloned Dolly the sheep. After initially accepting that this opened the door to the possibility of cloning a human, Wilmut and the Roslin Institute soon backed away from the subject, declaring that it is "scientifically possible, but ethically unacceptable." In the United States, the National Bioethics Advisory Board reached a similar conclusion and called for a moratorium on human cloning research. Other countries followed suit.

Enter Dr. Richard Seed. A Harvard Ph.D. in physics, Seed switched his interest to reproductive biology 20 years ago. His first project was to transfer embryos from prize cows to surrogate mothers. Then, in the 1980s, he founded a company, Fertility and Genetics, to use the same technique to transfer fertilized eggs from healthy women to women with fertility problems. Over the years Seed has published his biomedical research in such prestigious journals as *The Lancet* and *The Journal of the American Medical Association*. One paper in *JAMA* reported the birth of a healthy child by the technique, but it soon was eclipsed by the advent of in vitro fertilization.

With Wilmut and the Roslin Institute dissociating themselves from human cloning, Seed and his Human Cloning Foundation have become its champion. NBC's *Dateline* did a story on him, and he was featured in *Time* magazine and other major press venues when he announced his plan to clone himself. When asked for public comments, many other researchers say they find Seed an eccentric, who lacks both the expertise and the funds to successfully achieve that goal. However, the concept does have its supporters. The Human Cloning web site lists astrophysicist Stephen Hawking, Harvard Law Professor Laurence Tribe, and U.S. Senator Tom Harkin among those in favor of going forward with human cloning research. And at least one report in *The Guardian* claims Seed has raised $15 million and purchased a site on Hokkaido, Japan, where its facilities for treating infertile couples are scheduled to open in August, 1999 (2 December 1998). (Seed, however, declined to answer any questions on "seed money" in our interview.)

A professed Christian, opposed to almost all abortions and to assisted suicide, Seed explains to SKEPTIC how and why he believes cloning a human is just the first step on the road to his ultimate goal of "man becoming one with God" through unlimited rejuvenation.

Figure 18.0: Richard Seed

Miele: Dr. Seed, as Director of the Human Cloning Foundation (http://www.humancloning.org), your name has appeared in all the major media. Your training is as a physicist, not a biologist. Why have you become the point man for cloning a human being, rather than Watson or Crick or Ian Wilmut, who cloned Dolly the sheep? How did you get into cloning research and why? How are you qualified to lead this effort?

Seed: About 25 years ago I switched from physics to biology. I have a long string of short papers in biology that goes back to 1973. I started by studying embryo transfers in cows as a model for embryo transfers in humans. I spent about seven years on that research. I then spent about 12 years working on how to apply that technology to humans as a treatment for infertility. I was the lead researcher in the group that published the paper on the first human embryo transferred from one person to another.

Then I worked on a very closely related aspect of embryology—tolerance. We are tolerant of our own cells and we attack foreign cells. In terms of medicine and popular attention, the most relevant topic is organ rejection, but the body rejects any foreign body. Scientifically, the real question is self tolerance—how does the body distinguish between any foreign body and the self. Only a few scholars are really thinking about that question and even a smaller number are actually working on it. At present, we have no scientific theory of self tolerance. Once we really understand how that works we will solve the whole problem of organ rejection.

Miele: Why are people so afraid of cloning? There are the usual philosophical arguments in the reports of various commissions and in scholarly debates. But deep down inside, what do you think terrorizes so many people about the idea of cloning a human being?

Seed: Oh, that's very easy. In every media event involving a clone, the clone is a very bad human being. So the average person immediately associates cloning with evil. But this doesn't explain why some members of the intelligentsia are so hostile to the prospect. Because these critics are so highly educated their arguments are sophisticated enough that the average person doesn't care about them.

I remember one famous debate at the Royal Society in London in which the distinguished biologist J. B. S. Haldane spent a whole hour opposing the idea of human cloning because it was "contrary to the dignity of man."

That's a very sophisticated line of argument. But for the common man, including me, what is the "dignity of man"? I don't know what that means.

Miele: It's whatever the person making that argument says it means, right? Two reasons that spring to my mind are that cloning is associated on the one hand with the scandal surrounding David Rorvik's book, *In His Own Image* (NY: J. B. Lippincott, 1978), which claimed human cloning had been done 20 years ago, and on the other with Ira Levin's *The Boys From Brazil* (NY: Bantam, 1976, 1995 reprint), which links cloning to Nazi eugenics and racial hygiene. How important are those associations and how do you distance yourself from them, or do you?

Seed: As regards the general public, those are just two cases of the "bad media clone," but intellectuals don't even bring up the examples you mentioned.

Miele: But how do you, Dr. Richard Seed, respond if some critic says, "Boy, what you're suggesting sounds an awful lot like trying to breed a super man or a super race."

Seed: My response is that it's not relevant. You can use any technological means either for good ends or for evil ends. If you have the evil intent of making a race of slaves, why bother with clones? Dictatorships all around the world have done quite a good job of enslaving millions of people without bothering with cloning at all.

Miele: Rather than the old specter of imposed eugenics through roving government castration squads, isn't it more likely that the marketplace is going to demand all sorts of genetic technology, including cloning, just as it demands "miracle cures" for cancer, silicone breast implants, and designer drugs? People travel around the world and run all sorts of risks, including going to prison, just to try these and other means of self-improvement.

Seed: As you say, that's already happening. There is a demand for genetic technology and cloning today. As far as eugenics is concerned, I have a rather optimistic and favorable view of humans and human nature. Eugenics is already taking place. The number of Down's syndrome children in the United States, for example, has dropped by half in the last 10 years and is still going down.

Miele: Right, but that's an example of negative eugenics, trying to eliminate those genetic conditions we "don't like" (however that is defined).

But what about positive eugenics—the dream of breeding a super man or a super race?

Seed: We engage in that already as well. Children have their teeth straightened, women get breast implants, men get hair transplants, and so on.

Miele: But none of those things you just mentioned directly affects either our own genetic material or the germ line.

Seed: In my opinion you're making a fine distinction. We're already engaging in those activities. You're just describing one mechanism for doing so as opposed to other mechanisms. They are not substantially different. They are all designed to, and do, produce the same result.

Miele: Given that, aren't all those technologies, including genetic engineering and cloning, going to be available to the wealthy, but not for "the rest of us"?

Seed: Why is that always brought up? Why do you bring it up? Everything starts out being expensive and ends up being very cheap. I estimate by the 1000th clone the cost will be down to $10,000 apiece.

Miele: Then do you think Castro can save the faltering Cuban economy through its biotech sector by setting up a Club Med/Gen where visiting capitalists can get themselves cloned?

Seed: Wealthy capitalists aren't signing up to get cloned. I and the Human Cloning Project are fairly well known by now and wealthy capitalists aren't beating down our doors.

Miele: Who, then, wants to be cloned?

Seed: I already have between 200 and 300 infertile couples that have expressed an interest. They are ordinary, middle-class people, where one of the partners is sterile.

Miele: And what is their reason for wanting to try an experimental procedure like cloning rather than the readily available option of adoption?

Seed: They want to be able to transmit their own genes.

Miele: So are you simply advocating that we proceed with research on human cloning, or are you and/or your team actively engaged in trying to perform that cloning?

Seed: Both. Our team is doing research, and we encourage others to do research as well.

Miele: Can you give us a progress report? How close are you to cloning a human being?

Seed: I have to decline to answer that. I am no longer in a position where it can be discussed.

Miele: And the reason for that is?

Seed: Well, if you can't figure it out, then there is no reason. The fewer people know what I'm doing, the better off we are.

Miele: Then let me suggest a reason. The report to the president, *Cloning Human Beings* (Rockville, MD: National Bioethics Advisory Commission, 1997) stated, "at this time it is morally unacceptable for anyone in the public or private sector, whether in a research or clinical setting, to attempt to create a child using somatic nuclear transfer" (Chapter 6, p. 118). The report then called for Federal legislation to be enacted to prohibit anyone from attempting to do so. Are you working in opposition to those proposed Federal guidelines?

Seed: Absolutely. The subject was brought up in Congress. A number of bills were introduced in both the House and the Senate. All the bills were referred to committee. About nine months ago, the Senate held a vote to call several bills banning human cloning out of committee and they failed by a large margin to get the required majority. So, effectively, the Senate has spoken and said that it is not going to do anything to ban human cloning.

Miele: If there were a vote to ban human cloning, would you be willing to test the law by defying it, as did Dr. Kevorkian on the issue of assisted suicide?

Seed: Good Lord, no. Let's go on.

Miele: Who do you intend to clone?

Seed: First, let me say that the only legitimate criticism to human cloning that I recognize is the risk involved. In one debate I was accused of taking advantage of the plight of desperate infertile couples. I thought about it and it seemed there was some basis for making that charge. So I decided I would assume all the risk by cloning myself first with the help of my wife Gloria. If there was some abnormality, we would be sure to care for the child for the rest of its life.

Then someone claimed that it was all just a big ego trip on my part. Since there could appear to be some merit in that criticism, we decided to clone my wife, Gloria, first. So the plan is to have her carry her own clone in order to defuse those criticisms regarding risk and ego involvement.

Miele: You may be the only guy I know in the world who wants to have another copy of his wife. Most are desperate to get rid of the one that they already have.

Seed: Ha ha! Go ahead and print that Dr. Seed burst out laughing.

Miele: Okay! Cloning a human, or even a sheep for that matter, has PR value like, say, sending a man to the moon. But hasn't the history of the space program (and other research projects) shown that these big ticket, above-the-fold headline projects really impede the progress of science by diverting money, interest, and personnel from the less costly but more basic research that in the long run tells us more about how the universe works?

Seed: Understanding how the universe works or how life works? Understanding how the universe works basically has no payoff. Understanding how life works has the largest payoff of anything ever conceived. That's why I switched from physics to biology. What good does it do to know if the universe is straight or curved? Or to know whether it will expand continually or if it will eventually contract upon itself?

Knowing the answers to those questions won't do you and me one dime's worth of good. But if you understand life you could control life. Not only could you eliminate all existing diseases of the body and the brain, you could control the length of life and probably live indefinitely. Rejuvenation would produce the greatest return on investment of any project ever conceived. I estimate it will cost maybe one or two billion a year for 20 years. I think it's the number one science project in the world.

Miele: One of the arguments against cloning is that it would reduce genetic variability. Designer strains of corn produced in the green revolution are now being bred back to wild strains, precisely to increase their variability to reduce the susceptibility to disease. How do you respond?

Seed: That's a digression from cloning and I don't want to spend much time on it. Genetic technology allows us to manufacture artificial chromosomes and artificial genes. Instead of limited biodiversity, genetic technology has given us the means for unlimited biodiversity.

Miele: Then let's talk about the risks involved in cloning. A recent report in the prestigious British medical journal, *The Lancet*, stated that "cloning has a relatively high rate of late abortion and early postnatal death" with death rates of cloned animals around 50%. And that "cloning may pose an inherent risk to long-term health" and specifically advised that these figures "be taken into account in debates on the effective application of reproductive somatic cloning to human beings." What risk rate are you willing to accept? It took over 200 nuclear transfers to get just one cloned sheep.

Seed: Those are the only legitimate concerns that I have. I estimate with an expenditure of $20,000 to $50,000 you can reduce the probability of abnormalities in cloning to 1 in 10,000, maybe 1 in 50,000. Some goats were cloned recently and they were successful in 32 tries.

Miele: Is your wife willing to take 32 tries to get cloned?

Seed: No, no. You have to realize that most of them fail in the laboratory. The whole idea is to culture the embryos in the laboratory as long as you can. An embryo normally does not get from the oviduct into the uterus until Day 5, when it consists of about 150 cells. You should grow it up to that age before you place it in the uterus. Applying these methods has increased the pregnancy rate for in vitro fertilization to 50%.

Miele: What about the factor of aging? Aging of somatic cells wears down the ends of the chromosomes (telomeres).Wouldn't you be better off trying to clone someone who is 20 years old because you or your wife will have the accumulated effect of all those mutations?

Seed: There was a theory at one time that the telomeres got chopped off one segment at a time every time the chromosome divided. But now we see that there is a protein called telomerase that adds back on to the end of the chromosome anything that got chopped off. So it's uncertain whether the telomeres have anything to do with aging. In my opinion, they have nothing to do with it.

Miele: Also, what about the effect of the accumulated mutations in your or your wife's or any older person's cells? Wouldn't cloning be transferring these dangerous mutations into the germ line? Have you considered that?

Seed: The same aging effect would apply to sperm and egg cells, so it can't be too difficult a problem. It's certainly something to worry about, but it applies to the sperm and eggs in ordinary reproduction. Many mutations are also repaired. The nucleus of the cell has an inspection system that repairs mutations, but we don't know how effective it is and at what level.

Miele: Isn't the attempt to extend the much heralded Dolly experiment by cloning a human actually missing the flock for the sheep? The really amazing scientific fact in the Dolly case was that a differentiated somatic cell from a mature advanced mammal was in effect set back to zero and so reprogrammed to go through the entire developmental sequence. That tells us that the genetic program remains intact. Isn't understanding how the same genetic code translates into a brain cell in one case, a muscle cell in another, and a sperm or ovum in a third, the real mystery we're only beginning to discern?

Seed: Yes! We used to think that the program ran only one way and that only the germ line cells (sperm and eggs) were set up to repeat the process. Now it is clear that isn't so. Every single gene, and there may be 90,000 of them, has to have a closed control loop. The control loop is composed of a detector, an upper set point, a lower set point, a power source, and a switch to turn it on and off.

Further, this strongly suggests that you can reset the program to age 21 rather than age 1. Why go back to age zero or age 1? Why go back to the first cell division? Let's go back to age 21. Knowing what genes do is trivial. The Human Genome Project will tell us that. What we really need to know is the program that controls the genes and tells them to make brown hair or blond hair, or a muscle cell or a neuron—this is going to be at least 10 times as hard.

Miele: That's what I find more important scientifically than the pragmatic task of trying to make someone 21 years old again.

Seed: Right. The program is the most important aspect of DNA. Understanding the program is the number one scientific problem. Right now, the most important thing is to identity the role of each of the genes.

Miele: Then aren't we going to find out more about the program by performing lots of small and medium sized experiments with mice and sheep and others species, rather than going for the whole enchilada by cloning a human being?

Seed: Yes, yes, yes. But I've been saying for a year or two that cloning the first human will release a torrent of research regarding basic life processes— both the program and the genes. The torrent has already started.

Miele: Then why clone a human?

Seed: Because it will push the process along at 10 to 100 times the present rate. All the basic questions and basic research programs will lead toward the issue of rejuvenation—the road to eternal life. The brain will be the most difficult organ to rejuvenate. It will have to be done over a couple of years, because you'd like your new cells to make the interconnections necessary to maintain a continuity of consciousness and knowledge. Everything in your body except the brain will be easy to rejuvenate. Most organs can be replaced in a week or two. But it will, in my opinion, take years to rejuvenate the brain. It may take 5,000 different chemicals, one to stimulate re-growth in your kidney, one to stimulate re-growth in your liver, and so on. Assuming 90,000 genes, rejuvenation may require 90,000 different chemicals.

Suppose it all works, 200 to 300 years from now. There will be unlimited biodiversity. Not only can you have your teeth straightened or breasts enlarged or knees replaced, but you could get four arms instead of two—two small ones to work at the computer and two big ones to do the heavy lifting. If you don't like having four arms, you will be able to go back to having only two arms. Knowledge is accelerating at an exponential rate. Try to predict the next 40 years based on the last 40 years. Then try to project 200 years in the future. It's nearly impossible.

Miele: This is my final question and it gets us into the far reaches of philosophy. Assuming unlimited rejuvenation, then something like eternal life becomes possible. Without death, does life have any meaning at all? Isn't it the very fact that we face death almost every moment that makes life so precious? If you know you're going to have it forever, what's the point? It's the uncertainty of life that makes it a challenge, that makes it so special.

Seed: That certainly is the number one challenge, and I wish to solve that problem. I want to eliminate death. All of this is conceptually possible. It may take a long, long time, but it is possible. The program has been turned back.

Miele: Unfortunately my deadline has not. Thank you. Thousands of clones of this interview are on their way to the newsstands and mailboxes.

ABOUT THE AUTHOR

SKEPTIC Magazine Senior Editor Frank Miele regularly interviews cutting-edge scholars and best-selling authors, and reviews books and articles in behavioral science, history, politics, and religion. He has been an invited speaker at the Skeptic Society's Cal Tech Lecture Series, the Foundation for the Future's Humanity 3000 Symposium, the Westermarck Society of Finland, the Institute of Ethnology and Anthropology of the Russian Academy of Science, the Max Planck Society of Germany, and the Ludwig Boltzmann Institute of Vienna. He has also been interviewed on National Public Radio's *Science Friday.*

In addition to *SKEPTIC*, Miele has also published in *Intelligence, The Human Ethnology Bulletin,* and *Population and Environment* (where he also serves as book review editor). His articles have appeared on the Skeptic Society (www.skeptic.com), the Human Behavior and Evolution Society (www.hbes.com), and other websites.

Miele's unique writing style, drawing upon his varied life experiences, combines a technical writer's accuracy, brevity, and clarity, a musician's car for catching the rhythm, style, and personality of his iconoclastic and best-selling interviewees, and the wry wit of a stand-up comic.

www.ingramcontent.com/pod-product-compliance
Lightning Source LLC
Chambersburg PA
CBHW031821170526
45157CB00001B/133